Google Cloud for Developers

Write, migrate, and extend your code by leveraging
Google Cloud

Hector Parra Martinez

BIRMINGHAM—MUMBAI

Google Cloud for Developers

Copyright © 2023 Packt Publishing

Group Product Manager: Mohd Riyan Khan

Publishing Product Manager: Niranjan Naikwadi

Senior Editor: Sayali Pingale

Technical Editor: Nithik Cheruvakodan

Copy Editor: Safis Editing

Project Coordinator: Ashwin Kharwa

Proofreader: Safis Editing

Indexer: Manju Arasan

Production Designer: Vijay Kamble

Marketing Coordinator: Agnes D'souza

First published: May 2023

Production reference: 1040523

Published by Packt Publishing Ltd.

Livery Place

35 Livery Street

Birmingham

B3 2PB, UK.

ISBN 978-1-83763-074-5

www.packtpub.com

To my loving wife, Eva Celada, for her patience during the endless nights and weekends I spent writing this book, and her tireless support during this and many other not-so-easy moments during the last 15 years. You are my light; thank you for making me smile and feel so fortunate every single day. I love you!

– Hector Parra Martinez

Foreword

Google Cloud was launched some years ago to make available for customers all around the world the same infrastructure that Google uses for its end user products, such as **Google Search** and **YouTube**. We combine our offering of massive amounts of computing power with a versatile portfolio of managed services and open APIs, which can help developers integrate advanced capabilities, such as artificial intelligence, into their workflows in a very easy, secure, and cost-effective way.

Google Cloud is also the greenest cloud. Google has been carbon neutral since 2007, but we aim to run on carbon-free energy 24*7 at all of our data centers by 2030. We are also helping organizations around the world to transition to more carbon-free and sustainable systems by sharing technology and methods and providing funding.

Many of the readers of this book may be developers working for companies that don't use a cloud provider. This book will help you understand what Google Cloud is, how it can help your organization during its digital transformation, and how you can get the most out of it as a developer, engineer, architect, or IT professional. Once you get rid of the burden of having an infrastructure to maintain and rebuild your processes for agility and efficiency, the tangible benefits of the transformation will be really surprising.

Combining open source technologies, such as Kubernetes, with our own implementations of Google Kubernetes Engine and Anthos, *Google Cloud for Developers* will show you how to architect and write code for cloud-native applications that can run on Google Cloud and how to take them to the next level by containerizing them and making them available to run either on-premises, on Google Cloud, or even on other public cloud providers, taking you on a journey toward hybrid and multi-cloud architectures.

Technical information is complemented in the book with a nice list of best practices and tips, as well as identifying pitfalls that often arise when an organization is migrating an application to Google Cloud. This is a great resource with a lot of practical information that can help you save lots of time.

I also like the fact that this book is written by a Googler who was new to Google Cloud five years ago but now develops solutions for the biggest advertisers in Spain and EMEA.

This shows that the learning curve is not so steep and that spending time studying the different alternatives that we offer to architect and build cloud applications can have a significant impact on your professional career.

If you are a developer looking for a powerful, secure, reliable, and sustainable cloud platform, Google Cloud will be a great option. And this book will provide you with the information you need to start building and deploying applications that can run wherever you want.

Isaac Hernández Vargas

Google Cloud Country Manager for Spain and Portugal

Contributors

About the author

Hector Parra Martinez has worked in corporate IT for more than 15 years, specializing in failure monitoring and the automatic recovery of applications, systems, and networks. In 2018, he joined Google as a customer solutions engineer, helping the largest customers in Spain and EMEA to make the most out of Google Cloud for their advanced marketing analytics and data activation projects using Google Ads and Google Marketing Platform.

Hector is a certified *Google Cloud Digital Leader* and co-leads Google's *Mind the Gap* program in Spain, created to encourage more young women to pursue science and engineering careers. In his spare time, Hector is a big fan of retro gaming, TV shows, and electronic music. He also loves traveling with his wife, Eva, and spending quality time with his big family, especially his five grandchildren and two niblings.

I want to thank Google for believing in this project and making it possible, especially Eric A. Brewer for his kind support and Priyanka Vergadia and Miguel Fernandes for their titanic efforts to make this a great book. My deepest gratitude also to the Packt team that made this book possible: Ashwin, Sayali, Niranjan, Nimisha, and Agnes, you are really amazing!

About the reviewers

Miguel Fernandes is a senior solutions engineer with more than 16 years of experience in IT. He helps companies navigate their digital transformation efforts, with the last five years dedicated to delivering cloud solutions. He is currently working on privacy challenges as a privacy solutions engineer at Google. Having received an electronics engineering degree from Universidad Simón Bolívar, an MBA from Universidad Carlos III de Madrid, and a Master's degree in telematics engineering from Universidad de Vigo, he is passionate about scaling solutions globally using cloud technology while working within the constantly evolving privacy landscape.

I'd like to thank my family for being understanding of the time and commitment it takes to be on top of all the technology changes we are experiencing every day. This sector involves constant learning and I'm grateful for their acceptance and encouragement. I've also been lucky enough to work with great colleagues, like Hector—their support has made the experience so rewarding and immensely fulfilling.

Priyanka Vergadia is an accomplished author and public speaker specializing in cloud technology. As a staff developer advocate at Google Cloud, she helps companies solve complex business challenges using cloud computing. Priyanka combines art and technology to make cloud computing approachable through engaging visual stories. She has authored a unique cloud book (*Visualizing Google Cloud*) and created popular content, including videos, comics, and blog posts. Her work has helped many cloud enthusiasts get started, learn the fundamentals, and achieve cloud certifications. Find her on the YouTube channel and at `thecloudgirl.dev`.

Table of Contents

Part 1: Foundations of Developing for Google Cloud

1

Choosing Google Cloud   3

2

Modern Software Development in Google Cloud   21

3

Starting to Develop on Google Cloud   41

Part 2: Basic Google Cloud Services for Developers

4

5

6

Running Containerized Applications with Google Kubernetes Engine 119

7

Managing the Hybrid Cloud with Anthos 149

Part 3: Extending Your Code – Using Google Cloud Services and Public APIs

8

9

10

Extending Applications with Google Cloud Machine Learning APIs 231

Part 4: Connecting the Dots –Building Hybrid Cloud Solutions That Can Run Anywhere

11

Architecture Patterns for Hybrid and Multi-Cloud Solutions   269

12

Practical Use Cases of Google Cloud in Real-World Scenarios   285

13

Migration Pitfalls, Best Practices, and Useful Tips   305

Preface

Public cloud providers offer unlimited resources and a pay-per-use model, which opens the way to a new era of programming, no longer restricted by the lack of resources or the use of data centers located far away from customers.

Google decided to offer its internal technology to public users in 2008, and that's how Google Cloud was born. This meant universal access to massive processing and computing resources, together with a complete set of tools and services exposed using public APIs. This allows modern developers to easily extend their applications to benefit from the latest computing technologies and services, from modern infrastructure components to machine learning powered text, image, audio, and video analysis APIs.

I have written this book with a clear purpose in mind: to make it as easy as possible for developers to start writing, running, profiling, and troubleshooting their code in Google Cloud. But creating applications that run partially or totally in the cloud comes with its own list of challenges.

This book explains the pillars of digital transformation and how software development and project management have evolved in the last few years. The portfolio of services offered by Google Cloud has been constantly updated to remain aligned with the best practices of the industry and has become an invaluable tool for fast cloud application and service development, deployment, and migration.

While cloud computing is a trending topic, many organizations do not like to put all their eggs in the same basket. That is why this book also covers distinctive design patterns that combine on-premises and cloud computing resources to create hybrid and multi-cloud applications and services, making the most of each environment while diversifying the computing strategy.

This book also covers the most important parts of a migration to the cloud, from the initial questions to the long-term thinking process that will bring you closer to succeeding. A lot of real-world examples are included, together with lots of best practices and tips.

I have authored the book that I would have loved to have when I started my career, and I hope that it will be useful and will help you succeed. Thank you for reading it!

Who this book is for

This book has been written by a developer and is targeted at other developers and roles where writing code to run on Google Cloud is a requirement: cloud solution architects, engineers, and IT developers willing to bring their code to Google Cloud or start building it from scratch. Entrepreneurs in early-stage start-ups and IT professionals bringing their legacy servers and processes to Google Cloud will also benefit from this book.

What this book covers

Chapter 1, Choosing Google Cloud, begins with my own story as a developer and continues to explain how software development has evolved over the years. This chapter also covers the basics of digital transformation and why you should run your code in a public provider in general and on Google Cloud in particular.

Chapter 2, Modern Software Development in Google Cloud, begins by exploring the risks of traditional software development and how modern techniques mitigate these risks. The next part of the chapter covers how Google Cloud provides a set of tools and products that can be used to implement the mentioned modern techniques. The last part covers the different paths to migrate and write code that runs on Google Cloud.

Chapter 3, Starting to Develop on Google Cloud, starts by introducing the Google Cloud web console and then covers all the tools that can be used during the different phases of software development: Cloud Shell and its companion editor for writing code; it also mentions how to integrate Cloud Code in alternative **integrated development environments** (**IDEs**) such as Visual Studio Code, Cloud Logging and Cloud Monitoring for observability, and Cloud Trace and Cloud Profiler for troubleshooting.

Chapter 4, Running Serverless Code on Google Cloud – Part 1, covers the first two options for running serverless code on Google Cloud: Cloud Functions and App Engine, including how they work, what their requirements are, and how much they cost. The chapter also uses an example to show how we can use both options to run, test, and troubleshoot our code.

Chapter 5, Running Serverless Code on Google Cloud – Part 2, talks about Cloud Run, the third option available to run serverless code on Google Cloud, and explains the differences between the two environments that can be used. The example from the previous chapter is also implemented using containers, also explaining how to debug our code and how to estimate how much this option costs. The last part of the chapter is used to compare the three available options for serverless code, including some tricks to help you make the best choice.

Chapter 6, Running Containerized Applications with Google Kubernetes Engine, starts with an introduction to **Google Kubernetes Engine** (**GKE**), deep diving into the key topics, such as cluster and fleet management, security, monitoring, and cost optimization. The similarities and differences between GKE and Cloud Run are also explained, and tips are provided to help you decide when to use them. A hands-on example where a web application is containerized is also included.

Chapter 7, Managing the Hybrid Cloud with Anthos, starts by enumerating the key points to consider when choosing a cloud provider and how being able to work with different environments and providers simultaneously can be beneficial. Anthos is then introduced as a platform to easily manage hybrid and multi-cloud environments while providing unified management, security, and observability capabilities. After deep diving into Anthos components, concepts, and features, a hands-on example is included that can be deployed to either Google Cloud or Azure to better understand the benefits of Anthos.

Chapter 8, Making the Best of Google Cloud Networking, begins with a brief introduction to networking in Google Cloud, including how regions and zones work and how we can connect to our cloud resources. Next, some of the most important networking services available in Google Cloud are covered, including Cloud DNS, Load Balancing, Cloud Armor, and Cloud CDN. Finally, the two different Network Service Tiers are explained, and a sample architecture is used to showcase many of the network services and products discussed in this chapter.

Chapter 9, Time-Saving Google Cloud Services, this chapter covers some of the basic Google Cloud services that we can use to simplify our development process and our migrations to the cloud, including Cloud Storage to store our files, Cloud Tasks as a managed service for asynchronous task execution, Firestore in Datastore as a NoSQL database, Cloud Workflows to create end-to-end solutions, Pub/Sub for inter-component communication, Secret Manager to store our most sensitive data, and Cloud Scheduler to run our tasks and workflows exactly when we want. Finally, a practical exercise is included that combines most of these services.

Chapter 10, Extending Applications with Google Cloud Machine Learning APIs, explains how we can use Google's AI services and APIs to easily improve our own code. First, the differences between unstructured and structured data are explained, and then speech-to-text is covered as an example. Then, Cloud Translation is presented as a way to obtain final text files in the same language, and Cloud Natural Language is proposed as an interesting option to analyze these text files. In the next section, Cloud Vision and Cloud Video Intelligence are also presented as an alternative to help us understand the content of images and videos. Finally, a hands-on exercise is used to combine some of the mentioned services.

Chapter 11, Architecture Patterns for Hybrid and Multi-Cloud Solutions, starts by explaining the differences between hybrid and multi-cloud solutions and then justifies why these architectures make sense. Next, a list of some of the best practices to use when designing these kinds of solutions is provided. Then, hybrid and multi-cloud architecture patterns are divided into two different categories, and each of the design patterns is explained, including details such as the recommended network topology to use in each case.

Chapter 12, Practical Use Cases of Google Cloud in Real-World Scenarios, this chapter describes three very different scenarios and goes through the process of deciding which design patterns should be used to modernize and migrate each of them to Google Cloud. The key areas where we should focus our efforts are identified, together with the key decisions we need to take and the right sequence of actions to complete for these migrations to succeed.

Chapter 13, Migration Pitfalls, Best Practices, and Useful Tips, starts by identifying the most common pitfalls that happen while we move or modernize our applications to Google Cloud. Then, a list of best practices to bring our code to Google Cloud is discussed. Tips are included to help you overcome obstacles, handle delicate situations, which are quite usual in this kind of migration, and mitigate the complexity of this kind of process.

To get the most out of this book

You should understand the basics of writing, deploying, and running code. Basic knowledge of cloud services would be beneficial too, but a quick introduction is included for those who may be lacking it.

You will also need to have some familiarization with the Google Cloud web console and know how to use a Linux command shell. Also, since most of the examples are written in Python, knowing this programming language will make things much easier.

Software/hardware covered in the book	Operating system requirements
Python 3.x	Linux
Bash shell	Linux

If you are using the digital version of this book, we advise you to type the code yourself or access the code from the book's GitHub repository (a link is available in the next section). Doing so will help you avoid any potential errors related to the copying and pasting of code.

Download the example code files

You can download the example code files for this book from GitHub at `https://github.com/PacktPublishing/Google-Cloud-for-Developers`. If there's an update to the code, it will be updated in the GitHub repository.

We also have other code bundles from our rich catalog of books and videos available at `https://github.com/PacktPublishing/`. Check them out!

Download the color images

We also provide a PDF file that has color images of the screenshots and diagrams used in this book. You can download it here: `https://packt.link/d4IEw`.

Conventions used

There are a number of text conventions used throughout this book.

`Code in text`: Indicates code words in text, database table names, folder names, filenames, file extensions, pathnames, dummy URLs, user input, and Twitter handles. Here is an example: "Now we are ready for testing, and the four files have been copied to the same working directory: `app.yaml`, `favicon.ico`, `main.py`, and `requirements.txt`."

A block of code is set as follows:

```
DEFAULT_TEMPLATE = "english.html"
@app.route('/')
def get():
    template = request.args.get('template', DEFAULT_TEMPLATE)
    name = request.args.get('name', None)
    company = request.args.get('company', None)
    resume_html = return_resume(template, name, company)
    return resume_html


# This is only used when running locally. When running live,
# gunicorn runs the application.
if __name__ == '__main__':
    app.run(host='127.0.0.1', port=8080, debug=True)
```

Any command-line input or output is written as follows:

```
/home/<user>/.local/bin/gunicorn -b :8080 main:app &
```

Bold: Indicates a new term, an important word, or words that you see onscreen. For instance, words in menus or dialog boxes appear in **bold**. Here is an example: "The **Testing** tab can be useful for fast tests since it will help us quickly build a payload and trigger our Cloud Function, so we can then switch back to the **Logs** tab and check that everything works as expected."

> **Tips or important notes**
> Appear like this.

Share Your Thoughts

Once you've read *Google Cloud for Developers*, we'd love to hear your thoughts! Scan the QR code below to go straight to the Amazon review page for this book and share your feedback.

https://packt.link/r/1-837-63074-7

Your review is important to us and the tech community and will help us make sure we're delivering excellent quality content.

Download a free PDF copy of this book

Thanks for purchasing this book!

Do you like to read on the go but are unable to carry your print books everywhere?

Is your eBook purchase not compatible with the device of your choice?

Don't worry, now with every Packt book you get a DRM-free PDF version of that book at no cost.

Read anywhere, any place, on any device. Search, copy, and paste code from your favorite technical books directly into your application.

The perks don't stop there, you can get exclusive access to discounts, newsletters, and great free content in your inbox daily

Follow these simple steps to get the benefits:

1. Scan the QR code or visit the link below

https://packt.link/free-ebook/9781837630745

2. Submit your proof of purchase
3. That's it! We'll send your free PDF and other benefits to your email directly

Part 1: Foundations of Developing for Google Cloud

Public cloud providers have revolutionized software development, making resource restrictions and costly maintenance concepts of the past. Let's use this part of the book to discuss what these changes are, what they mean for an organization nowadays, and how corporate culture needs to evolve to embrace these changes.

We will also use this part to introduce Google Cloud as a platform that can help your organization implement all the new trends in software development, with a complete toolkit that will make writing, testing, deploying, and troubleshooting your code a much more enjoyable experience.

This part contains the following chapters:

- *Chapter 1, Choosing Google Cloud*
- *Chapter 2, Modern Software Development in Google Cloud*
- *Chapter 3, Starting to Develop on Google Cloud*

1
Choosing Google Cloud

I have written this book with a clear purpose in mind: to make it as easy as possible for developers to start writing, running, profiling, and troubleshooting their code in Google Cloud. Indeed, I'm a developer too, and I would love to share my story so I can better explain why I think, first, that you should be writing code on the cloud and, second, why, in my opinion, Google Cloud is the best option to do so nowadays.

We'll cover the following main topics in this chapter:

- My story as a developer
- Project management, Agile, DevOps, and SRE
- Introducing Digital Transformation
- Why should you run your code on the cloud?
- Introducing Google Cloud
- Why should you choose Google Cloud?

My story as a developer

I got my first computer in 1987 as a gift from my parents. I must admit that I chose an Amstrad CPC 464 because it came bundled with a few cassette tapes with games:

Figure 1.1 – Amstrad CPC 464 (author: Bill Bertram, source: `https://en.wikipedia.org/wiki/Amstrad_CPC#/media/File:Amstrad_CPC464.jpg`)

I used to play some of those games during my initial days with the computer, but they were unfortunately quite boring and primitive. One of them was called *Animal, Vegetal, Mineral* and it especially caught my attention. It was a very simple text game that tried to guess any animal, plant, or mineral that you could think of as soon as possible, by asking questions such as *is it a vegetable?*, *is it green?*, and *is it a lettuce?*

I have always been a very curious person and started to think how that game could work, and mentally pictured some kind of tree structure organizing all the information about potential answers so that the answers provided to the questions could help the code decide which leaf of that tree to traverse. As you can see, my mind was already a good fit for data structures, wasn't it?

The CPC 464 came with a small book, which was one of the reasons why I became a developer (I'll leave the other story about reverse engineering the protection scheme of a *Lemmings* game for another time). The title of the book in English was called *Basic Reference Manual for Programmers*. I loved reading at that age, indeed the worst punishment I could get was not being allowed to read at night, but this manual surprised and entertained me even more than the sci-fi classics I was used to reading because it opened the door to learning and fostered my digital creativity at the same time:

Figure 1.2 – CPC-464 Basic Reference Manual for Programmers

One step at a time, I was able to learn how to code in *Basic* on my own and implemented an even more primitive version of the game I mentioned earlier. But much more important than that, I loved the experience and decided at quite a young age that computers in general, and programming, in particular, would both be core parts of any professional career I would choose in the future.

Programming at that time was like writing with a pen on a blank piece of paper but using the bundled and primitive Line Editor application instead. In 1985, I just had that *Basic* manual to learn and get inspiration from. Forget about autocompletion, online help… and, oh my God, there was no Stack Overflow! Just take a look:

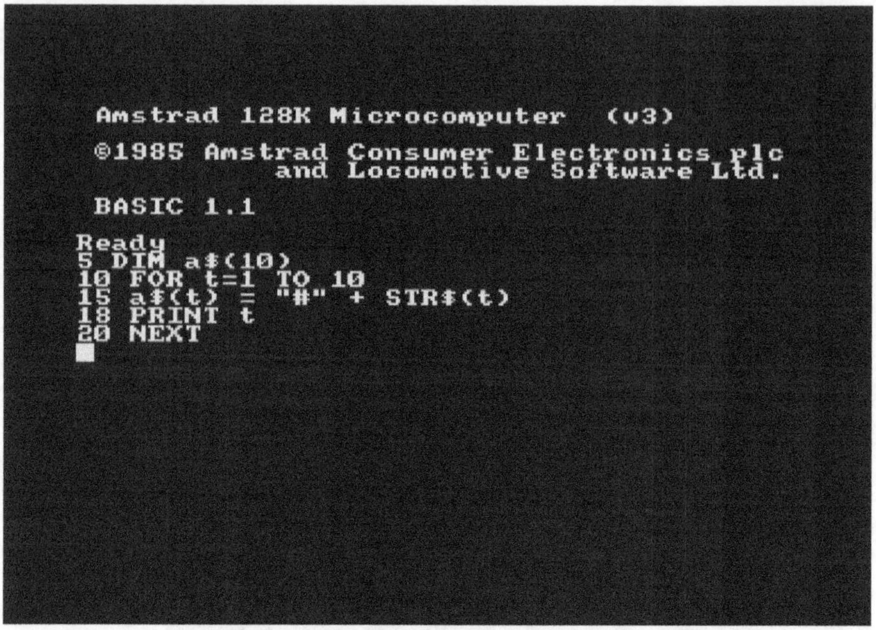

Figure 1.3 – Amstrad screen with Basic code

In the following years of my professional career, I had the chance to start working with Unix and Linux systems and started using **vi** and **emacs** as code editors for shell and Perl scripts, together with code written in C. I have always felt quite comfortable with the console, even for editing code, but I have to admit that graphical and integrated interfaces were game-changers.

A few years later, I moved to a pharma company where the presence of Windows systems was much more frequent. That meant meeting Microsoft's **Visual Studio** UI for the first time, and I must admit it was a very pleasant surprise. Coding in Visual Basic felt different when compared with my good old Amstrad CPC and exploring interface design was a very interesting experience. I also started using Notepad++ for my PHP, Perl, and bash scripts. It was great to see how all these applications added new features year after year.

But I still realized I had to create everything from scratch whenever I started working on a new project, and the time required to develop an application or service was just too long.

Fortunately, nowadays, there are a lot of different **integrated development environments (IDEs)** to choose from, with amazing features such as code completion, code control integration, online references, and samples, which make writing code a much more enjoyable experience. I love the idea of using web-based IDEs, which in my opinion make the experience comfortable. Being able to open a browser and have access to an IDE full of options is just amazing!

Visual Studio Code (`https://code.visualstudio.com/`) is one example that I use quite often when I develop applications for Google Cloud. Just compare the following screenshot with the previous one from the Amstrad CPC...

Figure 1.4 – Interface of Visual Studio Code

And we are getting closer and closer to code being automatically written by just providing a summary of what the piece of code should do (`https://www.forbes.com/sites/janakirammsv/2022/03/14/5-ai-tools-that-can-generate-code-to-help-programmers/`).

However, what I love about developing for the cloud is how easy it is to integrate external services that provide advanced features, or how accessible architecture patterns, reference implementations, or sample code are. Now, I can put together pieces of code during a single day of work that can do much more than what I could achieve years ago working for 2 weeks.

During my first few months at Google, I worked on a solution that analyzed display ads and let advertisers know what elements worked better. Knowing whether ads with a palm tree worked better than those with a swimming pool, or whether an image of a couple with a baby got more clicks than one with a group of friends was cool.

And implementing this application was reasonably easy thanks to what Google calls the **Cloud Vision API**, an AI-based service able to detect objects, text, and colors on an image. Imagine how long it would take me to develop this system on my own. I will admit it: I would never have been able to do it. But now, it just takes a few minutes to integrate the API of this service with my code.

And the same happens with other key services, such as storage, messaging queues, databases, and many others that we will cover later in this book. I can say loud and clear that Google Cloud has changed the way I understand and conceive software development. Not only can I develop applications much faster and much more securely but I can also deploy them much more comfortably and make them available to any amount of users worldwide. And all of this can be done from a browser running on my laptop.

Now, I can focus my time on innovative applications and use different components, which make use of cutting-edge technologies, to develop and deploy these applications in record time. And this is what Google Cloud can do for you, too.

So, long story short, after spending endless days fixing the effects of bad initial designs and upgrading hardware, operating systems, and applications, I realized that it would be much better if I could design and build resilient and distributed applications while reusing cutting-edge components and services, which would scale great and deploy faster, and try to decouple them from the hardware and the operating system.

But when we talk about developing, it's not just writing code but also getting to identify what problem or challenge needs to be solved and decide how we can provide a solution that works. And doing all of this properly is the real challenge.

In my case, I was lucky because, just before joining Google, my employer happened to start exploring ways to make projects shorter and more successful. That's how I became familiarized and started to use project management, Agile, DevOps, and, once I joined Google, **Site Reliability Engineering (SRE)** practices.

Let's discuss what these are and how they can help developers.

Project management, Agile, DevOps, and SRE

Organizations put a lot of effort into minimizing the waste of time and money in projects since both are frequently scarce. Being able to anticipate bottlenecks and blockers can help reduce the chances for a project to fail.

And it is here that project managers become the key players. Among other tasks, they are responsible for identifying stakeholders, dividing the work into tasks, assigning times to each activity, and following up to ensure that everything is completed on time.

Traditional project management used the so-called *waterfall methodology*, which divides a project into different steps that are completed in sequential order one after another: requirement gathering, design, implementation, testing, and maintenance.

However, there can be projects that may run for much longer than planned due to different reasons – for example, wrong or incomplete initial assessments leading to undetected dependencies, or never-ending tasks that block others.

Also, projects managed using waterfall methodologies are more rigid in terms of features. As these are defined in the initial phases, any changes due to unexpected reasons, such as a feature not being needed anymore or becoming obsolete, could derail the project.

Project management has evolved and one of the most common practices to reduce the risk of long delays is to split the project into different phases of incremental complexity, also known as *sprints*, while following an iterative approach instead of a linear one. These practices were introduced in more recent methodologies, such as *Agile*, which aim to speed up the progress of projects and offer tangible results as soon as possible.

In Agile and similar methodologies, a **Minimum Viable Product** (**MVP**) can be provided after completing just one or a few of the initial code sprints; then, the team will work on improving it using an iterative approach that adds new features and capabilities. It fixes any found bugs in each new sprint until the project meets all the requirements and is then considered to be finished.

The following diagram summarizes the different phases for each sprint:

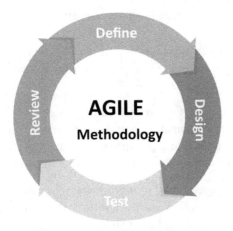

Figure 1.5 – Agile development phases

Agile is a project management methodology aimed at getting an MVP ready earlier, but it needs a compatible process on the development side to ensure agility. And here is where DevOps comes to the rescue.

DevOps is a set of practices that aims to increase the software delivery velocity, improve service reliability, and build shared ownership among software stakeholders. Many organizations use DevOps to complement Agile project management methodologies and reduce the lead time – that is, how long it takes for a team to go from committing code to having code successfully deployed and running in production:

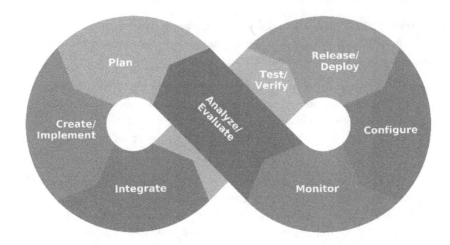

Figure 1.6 – DevOps development cycle (source: `https://nub8.net/wp-content/uploads/2019/12/Nub8-What-is-Devops-1-min.png`)

By implementing DevOps, you can improve a lot of your development metrics by increasing the speed of your deployments, reducing the number of errors in these deployments, and building security from the start.

These methodologies are very interesting for developers, but agility can only be achieved if the underlying infrastructure components are also compatible with fast deployments. For example, running short and fast sprints will not make sense at all in a platform where virtual machines are provided no earlier than 3 days after being requested and databases after 5 (and I have seen that happen, I promise).

An environment that can help you speed up all your processes is the best option, not only for developers but for everyone involved in IT projects. As we'll see shortly, the cloud is an extremely good option if you use Agile methodologies, are a big fan of code sprints, or want to implement DevOps in your organization.

And if DevOps helps automate deployments, SRE can also help in a later stage by automating all the manual tasks required to keep your environments up and running, such as those included as part of change management or incident response processes. And guess what – the cloud is a great place to implement SRE practices, too! To learn more about SRE, visit `https://sre.google/`.

If you are new to concepts such as Agile or DevOps, you may still be wasting a lot of your precious time as a developer doing the wrong kind of things. You should be spending most of your hours on innovating, thus contributing to the Digital Transformation of your team and the whole organization. We'll use the next section to explain what Digital Transformation means and why it is really important and should be appropriately prioritized if it hasn't been already.

Introducing Digital Transformation

I can imagine that many of you, while reading the first part of this introductory chapter, will have remembered your very own unpleasant experiences of working with infrastructure, applications, and architectures that started to grow and run out of resources due to a limited physical or virtual on-premises environment, a monolithic or overcomplicated initial design that made the application or service *die of success* after growing much more than expected and that you had to fix for good, or data split among so many databases in the organization that a minor update in the schema of a supposedly rarely used table broke most of the corporate applications.

The situations I just pictured are quite common among organizations that are still using an important amount of their IT time to decide where their infrastructure should run. And that's probably because they haven't completed their Digital Transformation yet. Even if you work for a start-up in its first stages, you may still be asking yourself these kinds of questions today. If that is the case, you should embrace the practices of digital transformation starting today.

The reason is that all these sadly common situations are incompatible with innovation. And IT professionals in organizations where innovation is constantly postponed because there are *other higher priorities* will become either outdated or burnt out, if not both, over time. If we combine this golden jail scenario with the burden of system and infrastructure migrations, there is a lot of precious time wasted on tasks that developers and engineers hate, and that don't add any value to the organization.

Let's say it loud and clear: if you want to innovate and if you want to be disruptive, you should focus your efforts on transforming or creating an organization where everyone can drive innovation. Otherwise, you will be wasting precious time and resources focusing on the wrong tasks.

Rob Enslin, former President of Global Customer Operations for Google Cloud, mentioned a few areas to focus on during a digital transformation process in a blog post from the Google Cloud website: `https://cloud.google.com/blog/topics/inside-google-cloud/innovation-in-the-era-of-the-transformation-cloud`. This list is, in my opinion, a very good summary of four of the main pillars of digital transformation, where organizations should put their efforts to free time and resources and be able to innovate more.

Let's comment on each of these pillars:

- *Accelerate the transformation, while also maintaining the freedom to adapt to market needs.* This is a very important point because while the digital transformation should happen in a reasonable amount of time, the process itself needs to be flexible too; otherwise, it may fail miserably if either the market or any other important external variable suddenly changes without prior notice. For example, during the pandemic, many companies were forced to speed up their digital transformation, and those who were ready to provide remote working capabilities for their employees earlier suffered less from the effects of the lack of productivity during those months that all of us had to spend working from home.

- *Make every employee, from data scientists to sales associates, smarter with real-time data to make the best decisions.* First-party data is power; however, it is often split into silos across an organization. A digital transformation should break down these silos by centralizing, deduplicating, and consolidating all data sources so that all the information is available to all members of the organization together with real-time insights that each department can use to make their own informed strategical decisions.

- *Bring people together and enable them to communicate, collaborate, and share, even when they cannot meet in person.* After the pandemic, it's even more clear that physical distance should not be a stopper, and all cultural elements of the organization should be replicable for people working remotely too so that people can also collaborate and share comfortably when they are far away from each other. Consider this as flexibility seen from a very specific angle.

- *Protect everything that matters to your organization: your people, your customers, your data, your customer's data, and each transaction you undertake.* Security is more important than ever, especially now that companies are using the power of technology to provide better services, and it should be a key element in any modern company transformation plan. Your data is your treasure and, together with your intellectual property, it might be what differentiates your organization from the competition. But it is also your responsibility to keep all your data safe, even more so when it probably contains personal and private information about your customers.

Rob summarizes these four pillars into their corresponding objectives: *application and infrastructure modernization, data democratization, people connections*, and *trusted transactions*. Any organization able to meet these objectives will have much more time and resources to dedicate to innovation.

If you read the previous paragraph carefully, you will realize that we developers are the key players in each of the four pillars of Digital Transformation, one way or another. During the digital transformation of an organization, developers will be working hand in hand with engineers on application and infrastructure modernization, which should be achieved by simplifying monolithic architectures by splitting them into elastic microservices. These apps and services will be using data as an input, and probably also generating data and insights as an output in many of the cases, so they will benefit from both the data centralization and the democratization mentioned earlier, and code should become simpler once data is easier to access.

And being connected to the rest of the team will also be important to make sure that our code meets everyone's needs. If we work using sprints, we need to be aligned with the rest of the team, even if each of us is located in a different office, country, or even continent. Finally, security is the key to ensuring that our apps and services are safe to be used and that our customers trust us more than ever.

Designing a Digital Transformation plan is not easy, and that's why there are a lot of companies working to help others succeed on their transformation journey. Some companies can help you design and execute the plan, but many others have created platforms that can make things much easier.

Some years ago, tech giants had the idea of abstracting the infrastructure up to the point that the customer wanted, letting organizations focus on what they love to do: architect, write, and run modern applications, centralize their data, make the most out of it, and get people connected, all of it in a secured platform. And guess what – Google Cloud is one of them.

Why should you run your code on the cloud?

There are a few reasons why I would recommend developers run their code on the cloud.

First of all, let me say once again: if you are spending too much time setting up servers, installing operating systems, deploying patches, and performing migrations, then you simply deserve better. I've been there and I can't put into words how happy I felt after I left it behind. I used to dedicate 20% of my time (and much longer on specific occasions) to maintaining, troubleshooting, and supporting the operating system, applications, and database. Since I joined Google, I can use that extra time to learn, brainstorm innovative solutions with my team, or write better code. I also think that *code is poetry* (https://www.smashingmagazine.com/2010/05/the-poetics-of-coding/), so IMHO, inspiration arrives better when we have more time and less pressure.

Besides, most cloud-based services offer customizable infrastructure components, or at least different sizes, so you can still have a reasonable degree of control over where your code runs. In summary, running code on the cloud will provide you with more options and better performance and will allow you to focus your time on coding, not on other distracting tasks.

Also, a cloud provider has many data centers in different locations across the world. If the start-up or organization you work for is planning to grow and have customers in more than one market at some point, a single server or even a few servers in a single location may not be enough to offer decent-quality service. This is becoming more and more important as real-time services become predominant and latency has to remain low.

If you can anticipate that you may suffer a potential scaling or latency problem in the future, already being in the cloud can make things much easier if you need to replicate your architecture in another continent for local users. Having infrastructure located closer to users can also help you meet legal requirements, as some countries require data or processes to be located in-country.

And speaking about scaling, the pay-per-use model is reasonably compatible with organizations growing because you will use more resources as you make more business and generate more revenue. Besides, most cloud providers will offer bigger discounts as you increase your usage. And if you have very particular needs, you can use huge clusters for a few minutes or hours and pay a very reasonable price. Indeed, you can have thousands of servers at your command at a very affordable price, something that would be prohibitive in an on-premises data center.

If your infrastructure is affected by traffic peaks, the cloud is also your place to go. If you have a lot of visitors on Sundays, your website crashes during Black Friday, or your app is usually down during the Christmas holiday season because of seasonality peaks, you may have decided not to increase the resources available for your application or website because, during most of the year, it can handle the average demand. With a cloud provider, you can scale up your application automatically when there is a peak and you can do the contrary too – that is, you can scale it down while your customers are sleeping so that you can reduce your costs. You can also schedule some tasks to run when the data center has less workload and save more money. We will discuss all of these opportunities later in this book.

What if you want to implement Agile and DevOps practices? Cloud providers have very fast provisioning times, so you can deploy complex applications and the infrastructure associated with them, as it is no longer static, in a matter of minutes. And that makes a huge difference, which allows you to use that extra time for better testing or even to do more iterations, which in the end will translate into better code.

And regarding the everyday life of a developer, if you are worried because you may not be able to keep on using your favorite IDE or fear that latency while writing code could be a problem, or that processes might be more complicated, just give it a try – you will be delighted. Hosting your repository in the cloud should be easy and you will not notice the difference. And you can connect from anywhere, even while commuting back home if you realize that you forgot to submit a very important CL before leaving the office.

I hope that you have been convinced that running your code on the cloud is a great idea. Now, let me show you why I think that Google Cloud is the best cloud provider to do so.

Introducing Google Cloud

Cloud providers offer different infrastructure components and managed services on demand using a pay-per-use model so that you don't have to worry about migrations, updates, patches, and similar time thieves.

Google's specific vision is to run their customer's code (and any other stuff they want to bring over to the cloud) on the same infrastructure used by its well-known products with billions of users, such as Google Search, Gmail, Google Drive, and YouTube. Using these same services is a guarantee of scalability and reliability. And this is what Google calls **Google Cloud**, a public cloud provider that many companies choose for their digital transformation journey.

If you are new to Google Cloud or are unsure about the number of products it offers, then it's a perfect time to visit Google's **Developer cheat sheet** (`https://googlecloudcheatsheet.withgoogle.com/`) so that you can understand the real magnitude of this offering; there are hundreds of services, organized in different areas, which allow you to accomplish virtually any task on the cloud. When you load the cheat sheet, you will see all the products, along with their corresponding names and icons, organized in different areas with different colors. You can scroll down to see the whole list; putting your mouse cursor over any of the tiles will show a very short description of each product.

A zoomed-out view of the cheat sheet looks like this:

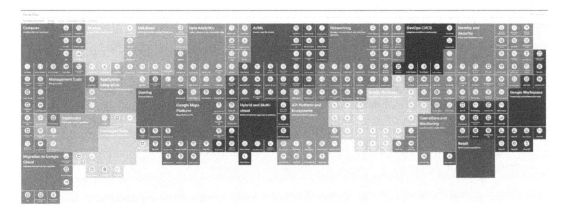

Figure 1.7 – Zoomed-out view of the Google Cloud cheat sheet (source: `https://googlecloudcheatsheet.withgoogle.com/`)

If you feel overwhelmed at this point, that's OK. I do too. This book is not aimed at going through that whole list, but to guide you on a quick and easy path to get you to write, run, and troubleshoot your code as easily as possible in Google Cloud. This book will cover those services directly or indirectly related to code development.

The product and service offerings of Google Cloud cover many different areas. I have selected just a few of the main services so that you can get a better idea of what I'm talking about:

- **Computing resources**: Virtual machines running in Google's Data Centers.

- **Serverless platforms**: Run your code without having to worry about the hardware or the operating system, including services such as Cloud Functions or App Engine.

- **Containerized applications**: You can use either Cloud Run or Kubernetes Engine.

- **Databases**: These offer all flavors: relational, NoSQL, document, serverless, and memory-based. They even offer managed instances of MySQL, PostgreSQL, and SQL Server and tools to easily migrate your database from Oracle, MySQL, and PostgreSQL to Cloud SQL.

- **Storage**: This is either for files or any kinds of objects and supports many different scenarios in terms of availability and retention.

- **Data analytics**: You can do this with a complete set of tools to help you ingest, process, and analyze all your data.

- **Artificial intelligence and machine learning**: These help turn your data into insights and generate models able to make predictions.

- **Networking**: This offers cloud and hybrid connectivity security solutions, together with load balancing and content distribution services, among many others.

- **Mobile platform**: This provides tools to help you make the most out of your mobile applications.

- **Hybrid and multi-cloud**: These options use Anthos to migrate, run, and operate your applications anywhere.

- **Migration tools**: These make it easier for you to move your stuff from an on-premises environment or other cloud providers.

But where are the services for developers? I'm a really bad guy and left them out of the previous list on purpose so that you didn't skip the rest. These are some of the development-related services that you can enjoy in Google Cloud, in addition to those already mentioned:

- Development tools and services, such as command-line tools and libraries, CloudShell, Cloud Source Repositories, Tools for PowerShell, Cloud Scheduler for task automation and management, Cloud Code, and IDE support to write, run and debug Kubernetes applications.

- DevOps continuous integration and continuous delivery (CI/CD) tools and services, allowing fast and safe code deployments with low error rates. Use Cloud Build for **CI/CD**, Artifact Registry to store build artifacts and dependencies, Google Cloud Deploy for fully managed purposes, Google Kubernetes Engine, Tekton for declaring CI/CD pipelines, and Cloud Deployment Manager to create and manage Google Cloud resources. Operations and monitoring tools and services are also provided, built to help you once your code is running in production. Log, Trace, Profile, and Debug can be used to troubleshoot any issue.

- A long list of public APIs provides a wide range of advanced features offered using a pay-per-use model that you can use to modernize your applications very quickly.

Indeed, the Google Cloud website has a page with a list of all the developer tools: `https://cloud.google.com/products/tools`.

Combine this a wide variety of managed services with the ability to connect your code to a huge API platform and ecosystem, allowing you to manage all your Google Cloud products and services. Besides, some of these Google Cloud APIs provide access to a set of machine learning that's pre-trained with advanced capabilities using a pay-per-use model, such as the following:

- **Vision API**: Able to identify objects, texts, colors, and faces in images and videos, and also flag explicit content

- **Speech-to-Text**: Used to transcribe audio into text (and vice versa) in more than 125 languages and variants

- **AutoML**: Allows you easily create, train, and productize custom machine learning models, even if you don't have any experience

- **Natural Language AI**: Allows you to analyze text, understand its structure and meaning, extract sentiment, and annotate it

- **Cloud Translation**: This is very useful for translating texts from one language into another
- **Dialogflow**: This can help you implement chat or voice conversation agents easily

These APIs can also help you simplify and modernize your applications by integrating the corresponding services to provide advanced capabilities with a few lines of code.

You can find the full list of APIs available in Google Cloud here: `https://cloud.google.com/apis`.

Technical documentation and videos are also available to help you solve some of the most common developer problems and use cases. You can read more about them here: `https://cloud.google.com/docs/get-started/common-developer-use-cases`.

Why should you choose Google Cloud?

There are many reasons why I would recommend you choose Google Cloud, not only to run your code but also to take your organization to the next level, because I picture us, developers, as main actors in any digital transformation process.

Summarizing all the topics previously covered in this chapter, these are the six key factors I would consider when choosing a cloud provider, in no particular order:

- Compatibility with Agile practices and app modernization
- Capabilities for data democratization
- People connections
- Protection and security
- Level of freedom and use of open software
- Cost-effectiveness

> **Note**
> Apart from my personal opinion, which I have shared during this chapter, to put together a more objective list of reasons why you should choose Google Cloud, let's review each of these factors and summarize all the information about these topics, all of which can be found on Google Cloud's website (`https://cloud.google.com/why-google-cloud`).

Thinking about app modernization and agility, Google Cloud is the first cloud provider to release a platform, Anthos (`https://cloud.google.com/anthos`), that empowers you to quickly build new apps and modernize existing ones to increase your agility and enjoy all the benefits of the multi-cloud. Also, the managed Kubernetes service seamlessly allows you to implement DevOps and **SRE** practices with cloud-native tools so that you can deploy your code with agility.

From the data democratization point of view, Google Cloud offers the ability to manage every stage of the data life cycle, whether running operational transactions, managing analytical applications across data warehouses and data lakes, or breaking down rich data-driven experiences. Besides, the key differentiator is that artificial intelligence/machine learning is a core component of the data cloud solution, which helps organizations not only build improved insights available to all members but also automate core business processes using data as the core.

Speaking about bringing people together, in Google Cloud, you can integrate video calling, email, chat, and document collaboration in one place with Google Workspace, which already connects more than 3 billion users. Google Workspace is built with a zero-trust approach and comes with enterprise-grade access management, data protection, encryption, and endpoint protections built in.

Protection is a key element of digital transformation, and Google Cloud can help you defend your data and apps against threats and fraudulent activity with the same security technology Google uses. Google keeps more people safe online than anyone else in the world: billions of users and millions of websites globally. Google pioneered the zero-trust model at the core of its services and operations and enables its customers to do the same. Besides, data is encrypted in transit between their facilities and at rest, ensuring that it can only be accessed by authorized roles and services with audited access to the encryption keys.

And if freedom is important for your organization, you should take into account that Google Cloud is the only cloud provider with a clear multi-cloud strategy. In Google Cloud, you can deploy and run each of your applications wherever you want: on-premises, on Google Cloud, or with other cloud providers. Google is also one of the largest contributors to the open source ecosystem, working with the open-source community to develop well-known open-source technologies such as Kubernetes, then roll these out as managed services in Google Cloud to give users maximum choice and increase their IT investments' longevity and survivability.

Another important point is that Google Cloud is open and standards-based and offers best-in-class integration with open-source standards and APIs, which ensures portability and extensibility to prevent lock-in, with easy interoperability with existing partner solutions and investments.

From a financial point of view, organizations can see significant savings when building on or migrating to a cloud-native architecture on Google Cloud. In addition, a reliable platform with 99.99% availability reduces risk and increases operational efficiency.

In summary, if you choose Google Cloud for your digital transformation, the result will be an organization and its workers being able to take advantage of all of the benefits of cloud computing to drive innovation.

Summary

I hope you are convinced about the benefits of choosing Google Cloud and how it is the best platform to help your organization simplify its development and infrastructure-related work to put more focus on innovation by completing your digital transformation, which will help you become much more competitive in your field.

In the next chapter, I will describe how developers work in legacy environments and highlight the differences in development workflows once you move to the cloud in general and Google Cloud in particular.

But before closing this chapter, and especially if you are new to the platform, before your development journey begins, I would recommend you make sure that someone in your organization takes care of building a proper Google Cloud Foundation by completing the 10-step checklist at `https://cloud.google.com/docs/enterprise/setup-checklist`.

And if you need more details about how to complete these steps, Packt has published an amazing book about it that I had the pleasure to review, called *The Ultimate Guide to Building a Cloud Foundation* (`https://www.amazon.com/Ultimate-Guide-Building-Google-Foundation/dp/1803240857`), so that you can start developing with peace of mind, knowing that all the basics have been taken care of.

Further reading

To learn more about the topics that were covered in this chapter, take a look at the following resources:

- *CPC Wiki page about the Amstrad CPC 464 and other family members* (`https://www.cpcwiki.eu/index.php/CPC_old_generation`).

- *The evolution to Integrated Development Environments (IDE)* (`https://www.computerworld.com/article/2468478/the-evolution-to-integrated-development-environments--ide-.html`)

- *DeepMind's AlphaCode AI writes code at a competitive level* (`https://techcrunch.com/2022/02/02/deepminds-alphacode-ai-writes-code-at-a-competitive-level/`)

- *The transition from Waterfall to Agile* (`https://chisellabs.com/blog/transition-from-waterfall-to-agile/`)

- *Agile Vs. DevOps: What's the difference?* (`https://www.guru99.com/agile-vs-devops.html`)

- *What is Digital Transformation?* (`https://cloud.google.com/learn/what-is-digital-transformation`)

- *Why Cloud Development Could (Finally) Become the New Standard* (`https://devspace.cloud/blog/2019/12/12/cloud-development-new-standard`)

2

Modern Software Development in Google Cloud

Development workflows have changed a lot in the last decades, as we started to discuss in the first chapter. In this one, we will set up the basis of what a developer does, so we can also discuss the potential associated risks and how modern development, especially on Google Cloud, can mitigate these risks and make the whole development experience much more enjoyable.

We will also introduce the set of tools that Google Cloud provides to help us developers be more productive. I will describe the different migration and development paths that you can use to get your code to run on Google Cloud, including how Anthos can be of help if you need to use hybrid or multi-cloud environments and take software modernization on them to the ultimate level.

We'll cover the following main topics in this chapter:

- What does a developer do?
- The risks of traditional software development
- How modern software development mitigates risks
- The benefits of implementing modern software development on Google Cloud
- Google Cloud toolbox for developers
- Migration and development paths to run your code on Google Cloud
- Managing hybrid and multi-cloud environments with Anthos

What does a developer do?

Answering this question is not easy, since there are many kinds of developers working in different areas, but let's try to summarize what they have in common.

Regarding skills, traditionally software developers were expected to have good **problem-solving skills** and be good both when working individually and when working as part of a team, with elevated levels of motivation and passion for their job. But what kinds of tasks does a developer take care of?

First, and obvious for sure, developers write or have written code to be run as a part of a script, application, or service at some point in their careers. They also probably (and hopefully) have written documentation for those pieces of code, so that others can know either how to extend them or at least how to use them. Some developers may also work on reviewing or documenting code written by others.

Code written by developers is usually the translation of requirements into an actual application, solution, module, or service, which could also be part of a project, and the work of a developer will often also involve tasks related to fixing detected bugs and supporting the users of our software.

Developers often work in teams, which involves some organizational work related to *who does what* and ensuring that the work of each of the different members does not interfere with the rest, using techniques such as **code control** tools and project management tasks to divide and assign the different pieces of work to the different members of the team.

While some of you reading this book may be working on developing small applications, or coding for start-ups in their first stages with not so many users yet, this book will also focus on the big scenarios because those start-ups you are at will get bigger very soon, and that's why we will also discuss complex and big applications that need to be ready to serve millions of users every day.

While this may seem an unnecessary generalization, it will help me to expose much more easily the potential risks that we may be facing as developers. As I will unceasingly repeat during the whole book, if you design and build an application that can handle heavy loads from the first iterations, you will save a lot of time and money when your user base starts to grow, especially if this happens suddenly or even unexpectedly. Thinking big from the beginning will never hurt a developer, but save a lot of time, money... and *pain*.

Picturing these big scenarios, I imagine a team of programmers and engineers working together on a big project to develop a complex application or service for millions of users. As you can imagine, there will be different areas of risk in such a scenario. Let's mention some of them and then discuss how modern development tries to mitigate them.

The risks of traditional software development

I worked in corporate IT for more than 20 years before I joined Google and dedicated most of my professional career during that time to developing code aimed at monitoring applications written by colleagues or sold by third parties.

Having witnessed, and also suffered, a lot of failures and issues during all those years, I will try to summarize all the potential risks of traditionally developed software that I can think of, in no special order.

Software bugs

Software bugs are undoubtedly the most frequent risk for developers. The quality of the code that we write does not entirely depend on our coding skills: a tight deadline, an excessive workload, or bad communication leading to a poor requirement-gathering phase may be among the varied reasons that can make it more probable for unexpected issues to be detected during the life cycle of our software, which we will need to promptly address.

In my opinion, **code reviews** are one of the most useful and interesting practices that can help reduce the number of bugs that reach production, while fostering a culture of collaboration and increasing awareness of what the rest of the team is working on. Integrating code reviews in our development cycle is vital to let bug detection happen before the software is available to the user.

However, a changing environment, especially sudden **scope changes**, which are common among developers and tend to make our lives more difficult, makes it difficult to get rid of bugs. So, it's much better to prevent as many bugs as possible from reaching production, while also accepting that some will reach our users undetected.

For this reason, having a proper process ready for the early detection of software bugs, fast mitigation by writing and testing patches, and quick deployment of the required fixes will help us deal with these bugs as soon as possible before they cause more harmful effects.

Slow development

When we speak about traditional software development, we are usually referring to **monolithic applications**, often written by a small team of people over the years and usually including thousands of lines of code in a single code base. This scenario will probably include more than one single point of failure because large code bases make poor-quality hacks more difficult to identify, and these can make the whole application easily crash if they cause memory leaks, or they can also be exploited by malicious intruders who detect them after reverse engineering the corresponding binaries.

Updating these kinds of applications is complicated because any changes in the code may affect multiple functionalities of the software, and the complexity of the source code can make it difficult to do proper testing.

Once we get the changes approved, we have another situation to deal with: deploying changes in monolithic applications automatically implies **application downtimes** when we perform the actual upgrades. If these upgrades are bundled with long data update operations, the downtimes can be significant.

Resource exhaustion

I already mentioned this risk in the first chapter and will also include it in this list, since it is usually the main reason for *our development headaches* due to applications not having enough memory available to run properly.

This risk is usually associated with monolithic applications and infrastructure availability issues, together with limited technical resources, all of them quite common in legacy environments and traditional software development processes. Apps developed in this scenario are usually not designed with built-in resilience and often crash for good when they run out of resources.

Lack of resiliency and fault tolerance

Our code should be able to handle virtually any potential issues that may occur. For example, if there is a power blackout or if a database starts to fail for a relatively prolonged period, we, as developers, should be able to design plans for keeping the application running while also keeping the **integrity** of our data safe.

Practices such as making updates in a transactional way or ensuring that there is **consistency** in every operation that we try to run can help mitigate this situation. We should only mark each of these operations as completed once we have verified that all the underlying tasks have been properly executed with a successful result.

This will be the only way to guarantee that we can not only recover from disasters but also avoid any catastrophic effects they could have on our applications and services.

Talking about disasters, running **disaster simulations** is a remarkably interesting way to be prepared for the unexpected. The list of **unit tests** that we may put together as developers to verify the successful execution of our code and all its functionalities will often not consider the potential effects of a rack exploding, a cable being cut by accident, or consecutive power outages. These and many other uncommon situations should be simulated periodically to prepare our solutions for the unexpected, using an approach that is compared with *the random chaos that a monkey could wreck while wandering around a lab or a data center*.

Indeed, using a **simian army** approach has worked well for Netflix, as you can read in this post in their technical blog (`https://netflixtechblog.com/the-netflix-simian-army-16e57fbab116`).

Failing to estimate usage patterns

Since most traditional software runs on limited resources, estimating **usage patterns** is key for allocating the right infrastructure size and avoiding application crashes. Once we understand how many users we are normally expecting, resources can be provisioned to ensure a decent service level for those users, adding an extra buffer of resources to support small peaks.

A widespread problem in these scenarios is that if we want to supply a satisfactory level of service during uncommon **seasonal peaks**, we will need to provision more infrastructure components, and most of these additional resources will be either unused or idling during the rest of the time, which is a total waste of money.

For this reason, these kinds of applications should be built to handle peaks by temporally dropping users in excess with a nice *Please try again in a few minutes* message. Thus, we prevent a complete crash that could be catastrophic if it happens at the busiest time of the day or the year.

Lack of proper monitoring and risk management

While the main goal for developers is writing code to build applications and services, additional time should also be spent on finding weak spots and potential risks in our software and building a proper internal and external monitoring plan that can check the status of these services by combining active and passive techniques.

During my career, I have seen big architectures not working because the SSL certificate of a non-critical web server had expired and services not working for days because a server had been shut down by accident without anyone noticing.

These are extreme examples, but they show the benefits of internal and external automated tests that simulate the behavior of our users and can be combined with log analysis and process and service checks, to verify the end-to-end availability of a service.

Unclear priorities, accountability, and ownership

One of the key factors to ensure fast and proper bug and disaster management is being able to understand who is responsible for each line of the code, especially when we are working in a team. Each component, service, or application should have a clearly identifiable author and a person responsible for issues detected within that component.

They may not be the same person in all cases, and that's why having a proper process in place to understand *ownership* and *accountability* will expedite the start of any recovery actions required and ensure that high-priority bugs are addressed as soon as possible.

It is also important to effectively manage **employee turnovers**, making sure that when a developer is about to leave the company, there is a proper handover process where one or more members of the team are involved and ready to inherit those responsibilities. It is also important that the **ownership information database** is updated accordingly so that the rest of the people in the organization are aware of these changes in the future.

Finally, the team should be aligned during the whole development process, sharing common priorities and practices, so that each team member is adding value instead of interfering with the work of the others. This is even more important if there are multiple teams collaborating to create a solution, with some of them even probably outsourced. In this kind of scenario, it is key to agree on common

design patterns, methodologies, and practices, so that processes such as **integration**, testing, and maintenance can be simplified, especially if the same team will be responsible for the whole code base in any of those processes.

Security approach

Traditional software development tends to build security as a separate layer built on top of applications or services. Besides, security is often added once the code has already been completed. It is often not updated when any changes are deployed during an update.

This approach to security greatly increases the chances of security accidents because security is decoupled from the services that should be protected. More weak spots can be exploited in both parts of our architecture when they are isolated from each other.

Lost source code

I have worked with running applications whose source code had been lost years ago, with no options for potential improvements, since only the binary was available.

Situations such as this can throw overboard years of work and it may be impossible to rebuild those applications ever again. For this reason, a proper **code version control system** is a must-have for all developers and, used together with **regular backups stored in multiple locations**, can save us from living dramatic situations and disasters that could make the work of a few months suddenly disappear in front of our eyes.

Now that we have gone through the main risks associated with traditional software development, let's take a look at how modern development practices can mitigate them.

How modern software development mitigates risks

Modern development workflows are significantly different from traditional ones. Even the required skills for developers have evolved, with a more prominent presence of **soft skills**, **data science**, or experience with software version control tools among the most wanted ones (https://www. botreetechnologies.com/blog/top-skills-software-development-companies- looking-for/).

Modern development has also ideated diverse ways of mitigating most of the risks mentioned in the previous section. Let's revisit and discuss how this can be done for each of them.

Software bugs

A few different practices can be combined to reduce the number of bugs and especially their potential to affect the availability of our application. First, code reviews should always happen following *the rule of "at least four eyes."* That is, at least the author and another developer should review the code before it is approved for submission to a code repository.

Code reviews should not be seen as *an exam or as a process where other people get to judge how good or bad we are as developers*. Instead, they are a fantastic opportunity to learn from other team members and write better code by iterating based on the feedback supplied by colleagues.

Reviewing other people's code also gives us the chance to see how they try to solve a problem, and honestly, I have learned a lot from this process and *eaten so many humble pies* after realizing that I still have so much to learn as a developer!

Resource exhaustion and slow development

Modern development tries to reduce resource exhaustion by dividing applications into multiple small, reusable, and independent components, known as **microservices**. These take care of a small part of the global functionality provided and are isolated from each other in terms of resource usage and communication using messages, in what is called an **event-driven** architecture pattern. These practices eliminate single points of failure and reduce the chances of a lack of resources in one of the components affecting some or all the other microservices in our application. However, microservices should be used only when they really make sense, as otherwise, they may add a lot of overhead and maintenance costs.

On top of this, **virtualization** and **containers** provide an easy way of supplying additional infrastructure resources to increase the available capacity before it gets exhausted. This, combined with the use of **load balancers**, can have two beneficial effects on our architectures:

- First, we can make our application or service *elastic* by increasing or reducing the size of the infrastructure depending on the number of users or the server loads, which will be frequently measured
- Second, we can redirect the traffic to infrastructure located in different data centers so that our users are always connected to the nearest data center, geographically speaking, minimizing **latency** and maximizing **throughput**

Looking at microservices from another angle, having independent services that communicate with each other allows different teams to develop each of these services; they just need to agree on the communication protocol for services to communicate and work together seamlessly. Each of these microservices could even be written using different **programming languages**, without this causing any issues.

All these situations make development much faster. Indeed, microservices architectures usually go hand in hand with **Agile** and **DevOps**, because services can be independently updated or improved, and deployments can be done very quickly without service interruptions and without affecting the availability of any of the other microservices in our architecture.

Lack of resiliency and fault tolerance

Resiliency is based on the *weakest link* principle, so a reasonable way to mitigate this problem is to use resilient infrastructure and communication protocols so that issues don't affect our service. For example, if I need an event message from one microservice to reach another, I can write a piece of resilient code with multiple delivery retries and **exponential backoff**, or I can use a third-party communication service that does it for me (as we will see later in the book) and queues my message, automatically retrying communication attempts if there are failures and informing me asynchronously when the delivery has been confirmed, even if this happened hours or days after the message was queued.

Failure to estimate usage patterns

Modern development techniques, where applications become elastic, mitigate potential usage estimation errors by measuring usage much more often. If our application checks whether the amount of allocated resources is enough every few minutes, and we can add or remove capacity very quickly when needed, then the application will become tolerant to peaks in the number of active users.

This does not only mean being able to handle the mentioned peaks but also being able to handle the opposite situation, reducing allocated resources when the number of users is lower than usual (for example, at night or during weekends or holidays). So, this feature, also known as zero scaling, can also help companies save a lot of money, as we will discuss in *Chapter 4, Running Serverless Code on Google Cloud – Part 1*.

Lack of proper monitoring and risk management

Implementing decent monitoring and **risk management** processes can become quite complex and the reason is simple: the number of different infrastructure components, applications, operating systems, and hardware used by an organization will grow exponentially with its size, and each infrastructure component will require a specific risk assessment and a customized monitoring plan.

The solution to minimizing these issues is to simplify or, even better, abstract our infrastructure, with the ideal picture being a data center where you just need to care about what you want to do (for instance, *run a Python script* and not worry about where it needs to run).

Once your view of the underlying infrastructure is homogeneous, risk management and monitoring processes will become much simpler. If the infrastructure is offered as a service, as it happens in the cloud, risk- and monitoring-related information may be offered as an additional feature of this service, making it even easier for you. **Simplification** is indeed a key step in the digital transformation process of any company.

Unclear priorities, accountability, and ownership

Mitigating the negative effects of the risks associated with these three topics can be done by taking different actions.

First, using a common set of practices can be achieved by using Agile and DevOps and agreeing on the duration of cycles, including fast deployment and common development rules for the team.

Also, using **code repositories** makes it easier to understand *who wrote what,* and having well-defined **escalation paths** can make it easier to establish ownership and decide *what to do next.*

Finally, a properly implemented **identity management system** will help avoid the potential harms of employee turnover by quickly finding which objects are owned by that employee and allowing them to easily transfer ownership, a process that should preferably happen before the employee actually leaves the company.

Security approach

Modern development is based on what is called **security by default**, which means that security is one more of the core topics that developers should always take into account when they design and build their applications. Integrating concepts such as **Zero Trust**, requiring users of an application to be constantly authenticated, authorized, and validated before they are allowed to use it, can make an enormous difference and reduce security-related incidents. In these situations, bypassing common security layers such as perimeter security, which this is not a replacement for but an addition to, will no longer be enough to gain access to the application or its data.

If the infrastructure that we use is also built using similar concepts at its core, then the global level of security will make it the perfect place to run our code.

After reviewing how modern software development practices mitigate risks, it's time to discuss how Google Cloud has mitigated these and other risks.

The benefits of implementing modern software development on Google Cloud

Google Cloud can help your organization minimize development risks and provide added benefits that will make developers more productive. Let's go through some of the features that make Google Cloud the best provider for modern development.

Built for agility, elasticity, and reliability

Cloud-supplied infrastructure components have four key features that differentiate them from traditional offerings: fast deployment, high availability, pay per use, and user abstraction.

This means that you can implement Agile and DevOps and be able to perform extremely fast deployments in your sprints, while you enjoy unprecedented levels of stability and reliability at extremely competitive prices and totally forget about administrative tasks at the infrastructure level.

You can also design elastic architectures where the size of allocated resources is dynamically adjusted depending on how many users are active at a given time, which optimizes costs and prevents resource exhaustion, giving you access to an unlimited pool of infrastructure components available using the pay-per-use model.

If you need a giant AI architecture to train your model just for a few minutes, you've got it. If you need to start 1,000 virtual machines for a one-off process lasting a couple of hours, you just request them, use them, and destroy them. This opens the door to a totally different way of understanding and using infrastructure, putting a virtually unlimited pool of resources at your fingertips, and letting you use it for as long as you need at a very reasonable price.

The different options available to run your code using a serverless service, from using containers to the **Function as a Service (FaaS)** model or virtual machines, also make it easier than ever to implement event-driven microservices and add support for service orchestration and choreography. We will discuss and compare all these services and options later in the book.

Once your application is deployed, there is still a lot to offer: the operations suite (`https://cloud.google.com/products/operations`), formerly known as Stackdriver, includes products that will allow you to aggregate logs from all the applications and services that you use on Google Cloud, and adds monitoring, tracing, debugging and profiling capabilities. We will talk about these services in the next chapter.

And if you thought that running your code on Google Cloud was comparable to putting it into a black box, you couldn't be more wrong: this suite will allow you to deep dive into both your applications and the components and services provided by Google and identify bottlenecks and issues, optimize your code, and improve your architecture. This is a gold mine for **Site Reliability Engineers (SREs)**, IT operations teams, and even for troubleshooting and improving your DevOps processes.

Even if you have public-facing APIs, Google Cloud has your back thanks to API Gateway (`https://cloud.google.com/api-gateway`) and especially Apigee (`https://cloud.google.com/apigee`), an amazing tool that helps you design, secure, analyze, and scale APIs anywhere with visibility and control.

And you don't need to care about patching, maintenance, and similar tasks; just request the infrastructure component that you need using either the UI or the API and destroy it when you no longer need it, since many of the infrastructure components used in the cloud architectures will be **ephemeral**, as we will discuss later in this book.

Talking about reliability, Google Cloud has 103 zones in 34 regions, which means having data centers available all around the world, allowing you to replicate your architecture across countries and even continents, providing the lowest latency and the highest throughput to your customers, while allowing you to implement **high availability** for your applications and services. And finally, let me remind you that Google Cloud provides a platform with 99.99% availability, which reduces risks and maximizes operational reliability, efficiency, and resiliency.

Security at the core

Google Cloud has security built in by default in its infrastructure offering, with some of the key security features including layered security and data encryption, together with extensive hardening, also for network communications, and a state-of-the-art process and team to respond to any detected threats.

As mentioned on the *Google Cloud Security* page of the official Google Cloud website (`https://cloud.google.com/security`), Google Cloud provides a secure-by-design infrastructure with built-in protection and a *Zero Trust* model that builds security through progressive layers, delivering true defense in-depth at scale. Data is encrypted by default, at rest and in transit, ensuring that it can only be accessed by authorized roles and services, and with audited access to the encryption keys.

Access to sensitive data is protected by advanced tools such as phishing-resistant security keys. Stored data is automatically encrypted at rest and distributed for availability and reliability, and no trust is assumed between services, using multiple mechanisms to set up and keep trust.

Built for developers

As I already mentioned, cloud providers are the best fit for teams using short cycles, speeding up development and deployment. This should be complemented with tools and services to ease the whole software development cycle, also including monitoring, debugging, profiling, and troubleshooting.

Google Cloud is the perfect choice for developers who use Agile and DevOps, or similar methodologies, allowing developers to minimize deployment and lead times, and making new features and bug fixes available much faster to the users of our applications.

A complete toolbox is available to help us implement these methodologies during all the stages of the development workflow. Let's take a look at this toolbox in the next section.

Google Cloud toolbox for developers

Google provides a toolbox aimed at improving productivity by providing advanced automation capabilities and centralizing information to make logging and troubleshooting tasks much easier.

> **Note**
>
> As I did in the first chapter, I will use the *Google Cloud developer tools* web page (`https://cloud.google.com/products/tools`) from Google Cloud as the official reference list to enumerate the different tools available to help us developers increase our productivity. I will also be adding my own opinions about them.
>
> The next chapter will cover how to use Google Cloud to write, deploy, run, monitor, enable logging on, troubleshoot, profile, and debug your code. So, this section plus the next chapter should provide you with a lot of information about the different tools that can help you succeed in your journey with Google Cloud and when and how to use each of them.

Let's take a look at the different tools available, divided into categories based on their main purpose:

- **Code**:

    - **Cloud Code** (`https://cloud.google.com/code`): This is a set of plugins for popular **Integrated Development Environments** (**IDEs**) that make it easier to create, deploy, and integrate applications with Google Cloud. **Remote debugging**, reduced context switching, and **Skaffold** integration are among my favorite features. Developers can keep on using the IDE of their choice (VSCode, IntelliJ, PyCharm, GoLand, WebStorm, or Cloud Shell Editor) and use Cloud Code to develop, deploy, and debug containerized applications on their Google Cloud projects, having a similar experience to when they are working locally. This tool is available free of charge.

    - Cloud **Software Development Kit** (**SDK**; `https://cloud.google.com/sdk`): Libraries and tools for interacting with Google Cloud products and services using client libraries for popular programming languages such as Java, Python, Node.js, Ruby, Go, .NET, and PHP. The SDK also includes the **Google Cloud Command-Line Interface** (**gcloud CLI**), a very useful tool to manage resources using the command line or to use with automation scripts. This tool is also available free of charge.

    - **Spring Framework on Google Cloud** (`https://spring.io/projects/spring-cloud-gcp`): Brings the Pivotal-developed (`https://pivotal.io/`) Spring Framework to the Google Cloud APIs to accomplish common tasks, such as exposing services and interacting with databases and messaging systems.

- **Build**

    - **Cloud Build** (`https://cloud.google.com/build`): A serverless **Continuous Integration and Continuous Delivery** (**CI/CD**) platform to build, test, and deploy your applications. It scales up and down with no need to pre-provision servers, just pay only for what you use, since each of your build steps is run in a Docker container. I especially like how it provides high-CPU virtual machines and a cache system to significantly reduce build times, together with its support for multi-cloud and built-in security, including vulnerability scans and the possibility to set up a secure CI/CD perimeter, blocking access to public IPs.

    - **Tekton** (`https://cloud.google.com/tekton`): A powerful yet flexible Kubernetes-native open source framework for creating CI/CD systems and helping you standardize your CI/CD tooling and processes across vendors, languages, and deployment environments. It works in both hybrid and multi-cloud environments and its goal is to let developers create and deploy immutable images, manage version control of infrastructure, or perform easier rollbacks. Tekton is a more powerful but also more complex option when compared with Jenkins.

    - **Jenkins on Google Cloud** (`https://cloud.google.com/architecture/jenkins-on-kubernetes-engine-tutorial`): A third option to help you set up a CI/CD pipeline with native Kubernetes support, GKE-based scaling and load balancing, and built-in CD best practices. Jenkins is more user-friendly but less powerful than Tekton.

- **Manage artifacts**

  - **Artifact Registry** (`https://cloud.google.com/artifact-registry`): This is an evolution of **Container Registry** and allows the creation of both regional and multi-regional repositories with granular IAM permissions and integration with either Cloud Build or directly with Google Kubernetes Engine, **App Engine**, and **Cloud Functions**. Some additional features include integrated security through binary authorization and vulnerability scanning, making Artifact Registry the best place for your organization to manage container images and language packages (such as Maven and npm) and set up automated pipelines.

- **Deploy**

  - **Google Cloud Deploy** (`https://cloud.google.com/deploy`): This is a managed service that automates the delivery of your applications to a series of target environments in a defined promotion sequence. When you want to deploy your updated application, you create a release, whose life cycle is managed by a delivery pipeline. A nice addition to your existing DevOps ecosystem, Cloud Deploy will allow you to create deployment pipelines for GKE and **Anthos** within minutes.

  - **Cloud Build** (`https://cloud.google.com/build`): This appears again in this list because it can also deploy your code using built-in integrations to Google Kubernetes Engine, App Engine, Cloud Functions, and **Firebase**, and supports complex pipeline creation with **Spinnaker**, adding an extra protection layer provided by Google Cloud. This is a very versatile tool to cover both the build and deployment phases of your development life cycle.

Now that we know which tools we can use in each phase of the development cycle, the last topic to discuss in this chapter is what the different options to migrate our applications and services to Google Cloud are and how to approach this process effectively.

Migration and development paths to run your code on Google Cloud

We have already discussed the potential risks of software development and how to mitigate them in modern environments, such as Google Cloud, and we also got familiar with the different tools that Google provides to help us become more productive as developers.

To complete the picture (and the chapter), let's discuss the different migration and development paths that you can use to get your code to run on Google Cloud, and explain how Anthos can help you in some cases during the process.

There will be a specific chapter dedicated to migrations at the end of this book, but I think it makes sense to introduce the topic in this chapter since this is one of the first decisions you need to take when you are starting to develop on Google Cloud.

Migration checklist

Before even starting to choose from the different options to migrate your application, there are a few questions that you should keep in mind.

Can this application run on the cloud?

This may sound like a dumb question at first, but if you are able to identify any limitations preventing your application from running on the cloud, and you are not able to find a valid solution or workaround, then you will be able to save a lot of precious time. Limitations preventing migrations are usually not technical, but come from areas such as licensing, compliance, or privacy, and they may be powerful blockers that cannot always be sorted out.

Some applications may also not be compatible with either **virtualization** or **containerization**, and that's another check that you should consider. When I say *this application* in the section title, please remember that your code may have **third-party dependencies** that you will need to test, too. Your code may be well prepared to run anywhere, but the libraries that you are using may not, and this will be the right time to study whether they can be replaced or whether you have found a solid blocker. So, you will not regret it if you invest some time to answer this question.

Is it worth migrating this application to the cloud now?

This is another apparently obvious question that many developers regret not having asked themselves at the right time. There are different points of view we answer this question from.

For example, how long are we expecting to have the application running once it is migrated? It may not make sense at all to have a team take 4 months to migrate an application and then receive a decommission request after a couple more months. Believe me, while this can always happen unexpectedly due to unforeseen circumstances, I have also seen this happening due to a lack of proper communication. So, please make sure that information flows correctly between areas in your organization before making big decisions regarding migration scheduling and prioritization.

Another possible angle to answer this question from is the complexity of the migration compared to the current number of users and how critical it is for the organization. Personally, I would hate to waste time migrating a legacy app that nobody used in the last 5 years, and that will never be used again. Having proper usage tracking metrics can help you save a lot of time and money. If an app must be migrated due to compliance reasons, at least knowing the metrics beforehand can help you choose the fastest migration path.

A third and final angle to answer the question is whether Google Cloud provides any service that may be used as a replacement for the whole application or service. If that is the case, migration could be considered the equivalent of *reinventing the wheel* and should be rejected.

Depending on the answers to these two questions, my suggestion is to go first with those applications or services for which you got two positive answers, prioritizing them based on how critical they are to the organization, and once you are done with the *quick-wins*, you can study the options for the most complex cases. You can also consider destinations such as Google Cloud Bare Metal Solution for Oracle (`https://cloud.google.com/bare-metal`) or even co-locate those servers that cannot be moved to the cloud, if your data center is being shut down as part of the migration and you need those applications to keep on running somewhere else.

Migrate or refactor?

Now that you have decided which applications are worth migrating, you will need a plan for each of them, and that's where you will choose which migration option is the best.

In most cases, you will be migrating monolithic applications or hybrid architectures running either on-premises or on other cloud providers. In general, there are three options that you can take when you decide to bring an application to the cloud:

- **Refactor your application**: This means fully rewriting or porting the application to the cloud, making use of all of its benefits to simplify the process while making the application better than ever. Depending on how you approach this process and the specifics of your application, this could be the most time-consuming of all options and it will rarely be the best choice.

- **Move and improve**: If you choose this option, you will be gradually migrating services, one at a time, to Google Cloud, and improving them using modern design patterns to decouple them from the monolithic core and implement them using an event-driven microservices architecture. This option may not be possible in all cases, especially with monolithic applications where partial migrations are not possible, but it can be an interesting option because it allows you to choose which services to migrate first, probably prioritizing those that are more problematic in the current architecture, making them benefit from all the features of the cloud first, and leaving the rest for a second phase. You could even decide to leave parts of your application in their current location (on-premises or on another cloud provider), thus creating a hybrid application.

- **Lift and shift**: Migrating an application *as is* is always an option. You can just clone your current setup in Google Cloud using either a virtual machine or a container, copy your binaries, libraries, and data files, and let the application run on the cloud in the same way as it did on-premises. There are a few scenarios where this could be the preferred option, such as virtualizing a legacy environment to enjoy the better scalability and pricing that Google Cloud offers, or for those applications that have very specific requirements to run properly and cannot be modernized, either because they were developed by a third party, because the original source code was lost, or because of legal requirements. In all these cases, there is a solution that will work for you.

While these three options will cover many of your use cases, there are still a few that are out of scope, especially those regarding applications that are either **hybrid** or **multi-cloud**. In these cases, you can try to migrate the part of the code running on-premises or on another cloud provider to Google Cloud and turn your application into a **cloud-native application**.

However, if you still need to keep some of your applications or services running either on-premises or on another cloud provider, Anthos is a Google Cloud service that can make it much easier to manage this kind of heterogeneous scenario. Let's explain what Anthos is and when and why it can make sense for you to use it.

Managing hybrid and multi-cloud environments with Anthos

Migrating your applications and services to Google Cloud is not always going to be possible and you may need to run some of them in different environments and platforms.

While this is technically possible, it can make things much more complicated, because you will need to manage different environments with their own capabilities, limitations, and requirements. And on top of that complexity, trying to have the same level of security, stability, privacy, orchestration, or compliance may just not be possible or it could be unmanageable for your development, IT support, and SRE teams.

This is where **Anthos** comes to your rescue. Anthos (`https://cloud.google.com/anthos`) offers a consistent development and operations experience for hybrid and multi-cloud landscapes. Consider Anthos as an abstraction provider for Kubernetes clusters, making it possible to deploy containers not only to run on Google Cloud, but also on-premises, on **Amazon Web Services** (**AWS**), or on **Microsoft Azure** (and the list keeps growing).

Use Anthos to deploy your clusters anywhere you want and manage them all in exactly the same way from your Google Cloud console. This sounds much better, doesn't it?

Anthos can be a very convenient tool when it comes to migrating your stuff to the cloud, especially if you are already using containers or if you confirm that your legacy applications can be containerized.

If you already run your applications using containers, you have three options to choose from. The first is to attach your existing Kubernetes clusters to Anthos (currently supporting Amazon EKS, Microsoft AKS, and OpenShift), and use some of the **centralization** and **observability** features that Anthos provides, while you keep on managing and updating your on-premises clusters manually.

The second option involves setting up Anthos locally on-premises, so you can either move your current clusters to your on-premises Anthos zone, or you may decide to move them to any of the other supported destination environments compatible with Anthos.

In all cases, you will be modernizing your applications and, once all your clusters are running on Anthos, you will be free to move the clusters to wherever you prefer at any time while still being able to centrally manage not only the containers but also the services they provide and the policies that you want to enforce in all of them.

This second approach is also valid if your legacy applications are not containerized, but can be because you can decide where to run each after containerizing them in Anthos, and still be able to manage your complete landscape no matter where each container runs.

The third option is valid for organizations where **VMware vSphere** is a corporate standard and also for those who are running their containerized applications on **bare-metal servers**.

If you are already running VMware vSphere, you can choose to run Anthos clusters on VMware and migrate your current virtual machines using **Migrate to Containers** (`https://cloud.google.com/migrate/containers/docs`). If you are running your containerized application on bare-metal servers, you can choose to install **Anthos clusters on bare metal** (`https://cloud.google.com/anthos/clusters/docs/bare-metal/latest`) and get rid of the **supervisor** for lower-than-ever latency and better performance.

In any of the last two scenarios mentioned, remember that you can also move your Anthos clusters to any other supported environment (Google Cloud, AWS, or Azure) whenever you want.

And once all your applications have been containerized and are managed by Anthos, you can create logical groups of clusters and containers located in different platforms. These groups are called **fleets** and allow you to group your resources, for example, depending on which environment they belong to. In this example, the group names would be *development, test, and production*. You can also set up different **regions** within a fleet to group resources by geographical location.

Once you create a fleet, you can apply changes to all members at the same time, which can save your administrators a lot of time and is perfect to integrate modern software practices such as CI/CD.

Some of the benefits that fleet members (`https://cloud.google.com/anthos/fleet-management/docs/fleet-concepts`) can provide include the following:

- Form, monitor, and manage a **service mesh** using **Anthos Service Mesh**
- Use **common workload identity pools** to authenticate and authorize workloads uniformly within a service mesh and to external services
- **Anthos Config Management** can be used to apply policy and configuration changes and is fully compatible with core Kubernetes concepts, such as namespaces, labels, and annotations
- Customize load balancing destinations using **Multi Cluster Ingress**
- Use **Cloud Run for Anthos** to enjoy all the benefits of **Knative**

Anthos also introduces the concept of **sameness**, a normalization process where some Kubernetes objects such as namespaces with the same name in different clusters are treated as the same thing to make grouping and administering fleet resources even easier.

Before finishing this chapter, I would like to elaborate a bit more on the concept of a service mesh, since it combines many modern software development practices. A **service mesh** (`https://cloud.google.com/service-mesh/docs/overview`) provides a dedicated and uniform infrastructure layer for *managed, observable, and secure communication across services*. This sentence, taken from the Service Mesh documentation page linked previously, means the ultimate abstraction of the most common points of concern when running an application using hundreds of containerized microservices: monitoring, networking, and security.

A service mesh has a proxy instance, called a sidecar, which connects to each application container and obtains information that is then centralized and automatically updated when new instances of a microservice are created, offering a clear picture of what's currently going on in your application and enabling unprecedented levels of observability and security while making global management possible with ease. All these features are offered at the cost of proxying all service requests, but it can be a lifesaver when the number of different microservices to manage begins to increase exponentially.

As you can imagine, Service Mesh is a fantastic addition to Anthos and brings modern software development to the next level of abstraction. Combining the fleet management capabilities of Anthos with the global observability and security features that a service mesh provides, you can simplify your processes and use most of your time as a developer for innovation.

Summarizing this last section, Anthos allows your organization to abstract infrastructure from applications and provides a console where you can group resources logically and manage fleets no matter where each cluster is running (on-premises, on Google Cloud, on AWS, or on Azure). This makes administration much easier and gives you total freedom to run each of your containerized services wherever you want and to move them from one place to another at your will.

Anthos fleets can use Anthos Service Mesh to deploy a dedicated infrastructure layer that provides centralized capabilities for observability, security, and common management options for all microservices, making Anthos even more convenient for software developers

As you can imagine, this is the culmination of software development modernization and a perfect ending scenario, from a software development process point of view, for the digital transformation of any organization.

Summary

In this chapter, we described what a developer does, according to traditional development workflows, the associated risks, and how modern development workflows in general and cloud development mitigate or get rid of those risks. Then, we enumerated the benefits of developing on Google Cloud and introduced the different elements of the toolbox that the platform offers to help professional developers like you and me be more productive. Finally, we described the different migration and development paths that you can take when you start developing or migrating an existing application to Google Cloud, remarking on how Anthos can help you build and manage hybrid or multicloud environments and take software modernization to the ultimate level.

The next chapter will focus on how you can use Google Cloud to write code, deploy and run it, set up logging, and monitor, profile, and troubleshoot your code, proving that Google Cloud is an amazing platform to cover all your needs as a developer.

Further reading

To learn more about the topics that were covered in this chapter, take a look at the following resources:

- *What does a Software Developer Do?* (https://www.rasmussen.edu/degrees/technology/blog/what-does-software-developer-do/)

- *12 Risks in Software Development* (https://www.indeed.com/career-advice/career-development/risks-in-software-development)

- *7 Advantages of Cloud Computing That Developers Can Benefit From* (https://datafloq.com/read/7-advantages-cloud-computing-that-developers-can-benefit-from/)

- *Use fleet Workload Identity* (https://cloud.google.com/anthos/fleet-management/docs/use-workload-identity)

- *Choosing between Cloud Run and Cloud Run for Anthos* (https://cloud.google.com/anthos/run/docs/choosing-a-platform)

3

Starting to Develop on Google Cloud

As we mentioned in the previous chapter, Google Cloud provides a set of tools to help us developers improve our productivity. This chapter will focus on introducing and describing the key features of these tools, so you can properly set up your development environment, while the next chapters will focus on examples and tips on how to make the most out of each tool when you run your code on Google Cloud. If you have previous experience with Google Cloud, you may want to skim-read or fully skip this chapter and go straight to the next one for some serverless action.

We'll cover the following main topics in this chapter:

- The first steps with the Google Cloud console
- Introducing Cloud Shell
- Writing code for Google Cloud using Cloud Shell Editor
- Writing code for Google Cloud using Visual Studio Code
- Setting up Cloud Logging
- Monitoring the execution of your code
- Troubleshooting by debugging, tracing, and profiling your code
- Appendix – testing your code on Google Cloud

The first steps with the Google Cloud console

Since some of you may be new to Google Cloud, I decided to include a very brief introduction to the Google Cloud console, so you can easily find everything in its **User Interface** (**UI**).

> **Note**
>
> Google Cloud is an ever-changing environment, and this is also applicable to its UI. When you read this book, the interface may not exactly match the screenshots used in the book, but they should still help you understand the concepts and find each element.

When we load the main screen of Google Cloud for an existing project (`https://console.cloud.google.com/`), we will either see a welcome page or the project dashboard, depending on our configuration, with quick access to the latest products that we recently used and their associated information, such as news, tutorials and documentation, platform status, API usage, billing, and monitoring, as shown in the following screenshot:

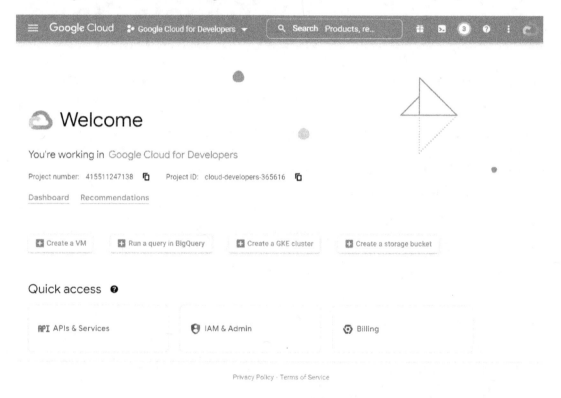

Figure 3.1 – Project dashboard screen in Google Cloud

There is also a blue ribbon at the top that provides easy access to all services and options:

Figure 3.2 – The blue ribbon at the top of the Google Cloud UI

Clicking on the hamburger icon (the first one with three horizontal lines) on the left side of the ribbon will open the services menu. Next to it, after the **Google Cloud** text, there is a drop-down menu to choose the project we will be working on. In the center of the ribbon, there is an omni-search box to easily find any product, resource, or documentation by just typing a few words describing what we are looking for.

The right side of the ribbon holds, from right to left, the account picture that we can click to switch accounts, an icon with three dots to access the preferences and settings menu, a question mark icon to open the help and support page, a bell icon to easily access pending notifications (or a number showing the number of unread notifications), an icon to open Cloud Shell, which I highlighted with a red box in the previous figure, and finally, a gift icon that will only appear if you are using free credits.

If you click on the Cloud Shell icon, you will see a message while your Cloud Shell **Virtual Machine (VM)** starts and, in a matter of seconds, you will see a Linux console on the bottom side of the screen, similar to the one you can see in this figure:

Figure 3.3 – Google console after opening Cloud Shell

Congratulations, you just opened Cloud Shell! Now, let's learn how to use it to write, run, and test code in combination with Cloud Shell Editor.

Introducing Cloud Shell

Cloud Shell (`https://cloud.google.com/shell`) is an online operations environment that you can access anywhere using a web browser and is offered at no added cost to Google Cloud customers. In other words, a Linux VM with persistent storage is provided for free to developers and administrators working on Google Cloud and its command-line console can be accessed from a web browser.

Cloud Shell is an online Linux terminal with a few preloaded utilities. This terminal, together with **Cloud Shell Editor**, can be used to write, deploy, run, and troubleshoot our applications on Google Cloud.

These are some of the key features of Cloud Shell that you should be aware of:

- **Pre-installed and up-to-date tools**: Cloud Shell includes many useful tools that you will often use, such as the **gcloud** command-line administration tool (`https://cloud.google.com/sdk/gcloud/`) and many other tools to help you manage software such as Kubernetes, Docker, Skaffold, minikube, MySQL, and so on. Of course, you can also install any other tools that you regularly use.

- **Persisting storage**: 5 GB of storage is provided and mounted in the home directory of your Cloud Shell VM, persisting between sessions. This makes Cloud Shell the perfect place to clone your repositories and then write, test, and deploy your code directly from your browser, and finally, commit changes back to the original repository.

- **Online code editor**: Write code using Cloud Shell Editor directly from your browser, a particularly useful feature for developers that deserves its very own section right after this one.

- **Cloud Shell VM and minikube Kubernetes emulator**: Run your code in Cloud Shell and test it in your browser before deploying it to production.

> **Tip**
>
> While Cloud Shell provides persistent storage and is a convenient tool for our development needs, *it should never be used as a replacement for a proper code repository*. Your storage will persist for a few days even if you don't use the Cloud console for a while, but after a few more days, you will receive an email warning you that your VM will be automatically shut down to save resources unless you use it again in a few days. *If this shutdown happens, your storage will be gone forever.* You can always click on the Cloud Shell icon again and the VM will start up, but its attached storage will now be empty. You have been warned!

Let's see how we can write code directly from the Cloud console.

Writing code for Google Cloud using Cloud Shell Editor

Cloud Shell Editor (`https://cloud.google.com/shell/docs/editor-overview`) is included as part of Cloud Shell and adds some interesting features to write code, letting us build, test, and deploy our code, all from our favorite browser.

Cloud Shell Editor is based on Theia (https://theia-ide.org/), an open, flexible, and extensible cloud and desktop IDE platform that supports languages such as Go, Python, Java, .NET Core, and Node.js. Among its features, we can enjoy syntax highlighting and context completions, linting, and code navigation, together with debugging capabilities.

The editor can be opened from Cloud Shell, using the button highlighted in the following figure:

Figure 3.4 – Details of the Open Editor button in Cloud Shell

But it isn't all good news...

> **Note**
>
> Cloud Shell has a *limited usage quota per week*, which also includes Cloud Shell Editor. Make sure to check your available quota so you don't run out of time while you still have pending tasks. If it is not enough, you can request a quota increase by contacting Cloud Customer Care (https://cloud.google.com/support).

Let's go through the features of Cloud Shell Editor in the following section.

Taking a look at the interface

In the top menu of Cloud Shell, you will see a button with a blue icon of a pencil and the **Open Editor** text (I highlighted it with a red box in *Figure 3.4*). Just click that button to open Cloud Shell Editor. After a few seconds to provision your editor instance, the screen will change, and you will see the editor:

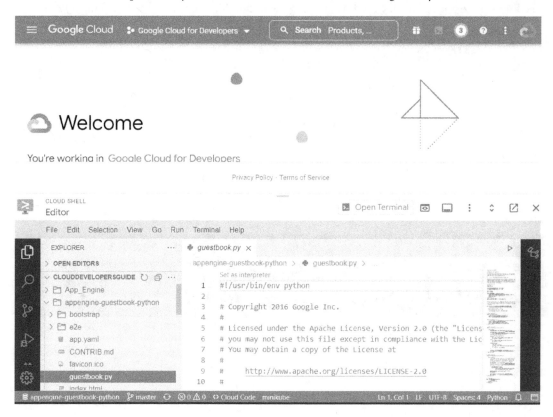

Figure 3.5 – The Google Shell Editor screen after clicking on the Open Editor button

The Cloud Shell Editor UI has a button with the **Open Terminal** text in blue letters to switch between the editor and the terminal, which, at least in my opinion, is not too comfortable for daily use. If you agree, there is a *maximize* icon to the right of the **Open Terminal** button with an arrow pointing to the top-right corner that you can click to open the editor in full screen in a new window, where the terminal will also be available at the bottom. This looks much better and the visual experience is quite similar to any other IDE:

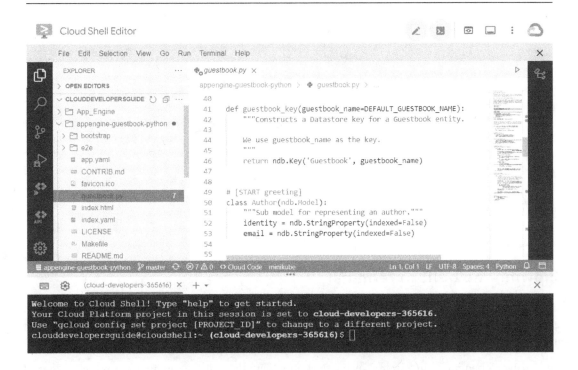

Figure 3.6 – Cloud Shell Editor in full-screen mode

> **Tip**
>
> If you find Cloud Shell Editor to be an interesting tool that you will use often, you can open it directly in full screen using this bookmark: `https://ide.cloud.google.com/`

The main screen in Cloud Shell Editor has a toolbar with icons at the top, a menu right below, an action bar on the left side, and two panels: a file browser on the left side and the main code editor in the middle. Finally, there is an outline bar on the right side.

The icons in the toolbar at the top-right side of the screen should already be quite familiar to you and can be used, respectively, to close the editor, open the terminal panel at the bottom, configure and launch the web preview, check your usage quota, access uploads, downloads, and other configurations (using the icon with three dots), and switch accounts.

The activity bar on the left side has a set of icons, each with its own purpose. I have included a description for the icons in the following figure next to them, so you can have an idea about the different functionalities that Cloud Shell Editor offers:

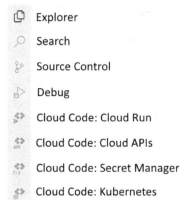

Figure 3.7 – The Cloud Shell Editor icon toolbar with descriptions

Showing the built-in terminal

The built-in terminal is a remarkably interesting feature since it allows you to use Cloud Shell without leaving the editor, so you can manage your cloud infrastructure, run and test your code, or use it for any other needs directly from Cloud Shell Editor.

You can even open multiple terminals using the button with the plus (+) sign in the terminal toolbar or using the **Terminal / New Terminal** menu item. The former will take you to a Google Cloud project selection screen before opening the terminal, which makes it possible to open a terminal directly connected to a different project, which can be useful to compare the configuration among different projects or other similar cases.

Uploading and downloading files

You will often need to upload or download files to/from the editor, and you can easily do this in two different ways:

- The first one is to right-click on an empty area of the file explorer panel to open a menu that includes one option to upload one or more files and another to download the current directory as a TAR file.

- The second way is to use the three-dots icon at the top-left side of the Cloud Shell Editor window, next to the thumbnail of your account image. Here, you will also see options for uploading or downloading files. I personally prefer to use this one for downloading files, because it will let you write the full path to the file or directory of your choice and will let you download it, which can be useful, for example, if you are using workspaces.

Editing and writing code

Cloud Shell Editor is compatible with the concept of **workspaces**, which means that all files used by a project are found below a specific root directory level (which is also the root level of the workspace) so that a download of that directory and all its subdirectories will contain all project files.

The benefit of opening a workspace instead of a single file is that you get instant access in the file panel to all project files, which is much more comfortable than having to use the *browse* dialog to locate each file that we need to open.

There are also more advanced workspace management options that you can read about on this page: `https://cloud.google.com/shell/docs/workspaces`

While you can open a file by clicking on it from the left-side file panel, you can also open it directly from the Cloud Shell terminal using a command like this:

```
cloudshell edit README-cloudshell.txt
```

Version control

Cloud Shell Editor supports accessing your existing Git repositories or even creating new ones using the **Source Control** button in the activity bar. You can also host your private repositories in Google Cloud using **Cloud Source Repositories** (`https://cloud.google.com/source-repositories/docs`). From that panel, you will also be able to see existing and staged changes and merge those changes.

Whenever you choose an action that has to do with your code repositories, Cloud Shell Editor will ask you to authenticate with your password, if you haven't done it recently. For this purpose, you should use a **Personal Access Token** (**PAT**), recommended by GitHub as a more secure alternative to standard passwords (`https://docs.github.com/en/github/authenticating-to-github/keeping-your-account-and-data-secure/creating-a-personal-access-token`). You can also turn on and configure **Git credentials helper** (`https://git-scm.com/book/en/v2/Git-Tools-Credential-Storage`) to make the use of PATs more comfortable by enabling caching of your PAT and increasing the time it is cached, so it's not constantly asking you to enter it again.

Once your setup is ready, you can clone a repository by going to **View | Command Palette** and running the `"Git: Clone"` command. You can also create a branch or switch to another one using the `"Git: Checkout"` command.

Now that everything is ready, it's time to code! At this point, the **Changes** section of the **Source Control** panel will display each of the files that have been changed, and opening each file will show you which lines have been changed but not committed yet. You can also click on a filename to display pending changes and decide whether to stage or discard them.

If you decide to stage your changes, the new **Staged Changes** section in the **Source Code** panel will allow you to click on a file and see a diff command view showing all the changes after comparing it with the earlier version.

At all times, an icon will show you which branch you are working on and a summary of the current status, and you can use the **Synchronize Changes** action from the **Command Palette** to push your local pending changes to the chosen branch and pull recent remote changes.

Finally, you can commit your changes using "Git: Clone" in the Command Palette.

Of course, if you feel more comfortable using a terminal for all Git-related commands, you can use Cloud Shell to clone your repository, add files, include a comment, and push the changes yourself, while using the editor just to write code.

Cloud Code support

Cloud Code is directly integrated into Cloud Shell Editor. The last four icons in the **Activity Bar** provide access to options for different Google Cloud services, Cloud Run, Cloud APIs, Secret Manager, and Kubernetes, as you can see in *Figure 3.7*. A clickable button in the status bar with the text **Cloud Code** can be used to open a menu with different options, as you can see in the following figure, where all UI elements related to Cloud Code have been highlighted in red:

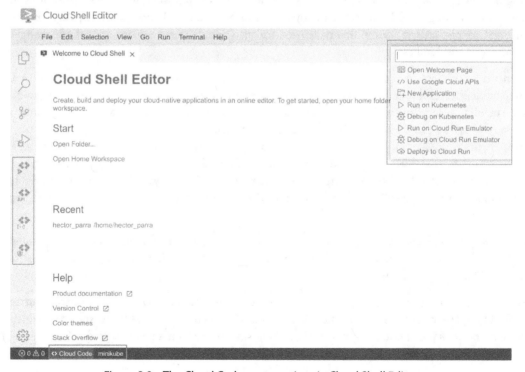

Figure 3.8 – The Cloud Code menu options in Cloud Shell Editor

Cloud Code includes sample applications that we can use as templates for faster prototyping and makes it easier to debug our code by including direct access to a **Cloud Run emulator**, particularly useful when testing containerized applications. All these menu options are available by clicking on the **Cloud Code** text in the status bar.

There are direct links for debugging an application, either on **Kubernetes** or on **Cloud Run**, together with a last option to deploy our code to Cloud Run once our testing is successfully completed.

Next to the **Cloud Code** button in the status bar, there is also a **minikube** button that allows you to instantly connect to a minikube cluster without leaving the IDE.

In the next chapter, we will see detailed examples of how to use Cloud Shell Editor in all the phases of our development workflow (write, test, deploy, run, and troubleshoot) for the different options available to run our code on Google Cloud.

Moving your code to a different IDE

Moving a project from Cloud Shell Editor to your favorite IDE is quite easy. The first step is to copy your workspace files somewhere else. You can either download your files using the **File | Download** menu or commit your latest changes to your code repository, using either the Cloud Shell Editor UI or some of the standard Git commands run from the terminal window.

Then, open your favorite IDE and either import the downloaded files, preferably by using a workspace-compatible IDE to make it easier and faster, or clone or import your repository using the UI of your chosen IDE. A third option would be to use Git command-line commands to clone the repository and make the files available in a local directory of your choice.

Once the code is available in your favorite IDE, you can set up and use Cloud Code to make your development tasks easier. Let's see an example with **Visual Studio Code (VS Code)**.

Writing code for Google Cloud using VS Code

Cloud Code is a set of plugins that provide support for different IDEs and makes it much easier to work with Kubernetes and Cloud Run. I have chosen VS Code as an example built on the open source Code-OSS, but you can follow a similar process with any of the other supported IDEs: IntelliJ, PyCharm, GoLand, WebStorm, and, as we have already seen earlier, Cloud Shell Editor.

In order to install VS Code, we should first make sure that all the prerequisites mentioned in the Google Cloud documentation (`https://cloud.google.com/code/docs/vscode/install`) are met:

- Install VS Code (`https://code.visualstudio.com/`)
- Install and configure the support for the languages that you will be using: Go (`https://marketplace.visualstudio.com/items?itemName=ms-vscode.Go`),

Python (https://marketplace.visualstudio.com/items?itemName=ms-python.python), Java (https://marketplace.visualstudio.com/items?itemName=vscjava.vscode-java-debug), and .NET (https://marketplace.visualstudio.com/items?itemName=ms-dotnettools.vscode-dotnet-pack)

- Install Git (https://git-scm.com/book/en/v2/Getting-Started-Installing-Git), which is required for copying code to and from your IDE

- You may also need to install the Docker client (https://docs.docker.com/install/#supported-platforms), unless you will use Cloud Build for building

I will be assuming that you already have access to a Google Cloud project with billing enabled.

Once the installation is complete, and after launching VS Code, you should see a screen similar to this one:

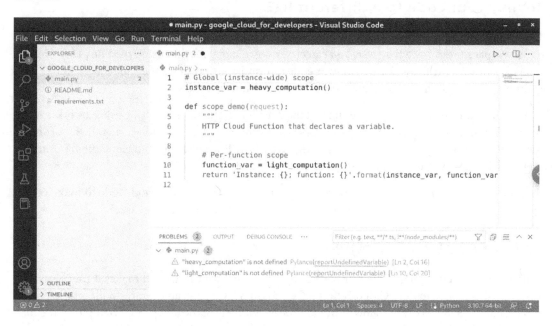

Figure 3.9 – The VS Code main screen

Installing the plugin

In order to install the plugin, just visit the **VS Code Marketplace page for Cloud Code** (https://marketplace.visualstudio.com/items?itemName=GoogleCloudTools.cloudcode) and click on the **Install** button. You may be asked to restart VS Code for the plugin to be enabled.

Now, you should be able to see the familiar Cloud Code icons that we mentioned for Cloud Shell Editor and can further customize your settings using the top-level menu **Code | Preferences | Settings | Extensions | Cloud Code**.

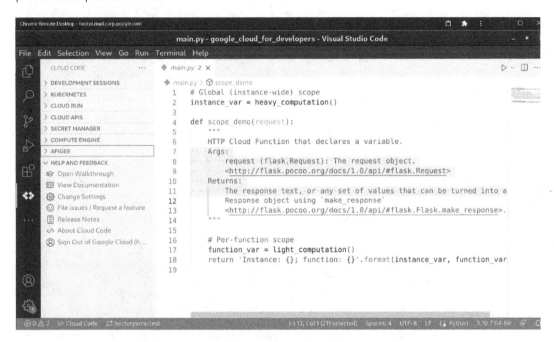

Figure 3.10 – The VS Code main screen after installing Cloud Code

Please notice that the plugin installs some dependencies automatically by default (`kubectl`, `Skaffold`, `minikube`, and the `gcloud` CLI command). If you prefer, you can disable this option by visiting **Manage | Settings** and setting **Autodependencies** to **Off**, and then proceed to install them manually at your convenience.

Now that we have discussed how to write code either in Cloud Shell Editor or in your favorite IDE (using VS Code as an example), we will start preparing our environment for testing our code.

Setting up Cloud Logging

One of the first tasks to complete is to properly set up Cloud Logging in our cloud project. This will make it possible for any service and application running on this project to send logging events that will be centralized in **Cloud Logging** (`https://cloud.google.com/logging`), where we will be able to filter and display a specific list of events whenever we need them.

Cloud Logging, part of the **Google Cloud operations suite** (`https://cloud.google.com/products/operations`), formerly known as Stackdriver, is a managed real-time logging service that allows us to store, search, and analyze our logs and set up alerts when certain events are received. Logs can be stored in regional log buckets, and inclusion and exclusion filters can be set up to decide which events we want to capture. Cloud Logging not only includes events generated by our code and by cloud platform components but also audit-related messages to help us gain visibility about who did what, when, and where within our cloud project.

You can quickly access Cloud Logging either by searching for `logging` in the search box at the top, or you can have it pinned in the list of services available by clicking the hamburger icon in the top-left corner of the screen.

Once you have accessed it, you should see a screen like this:

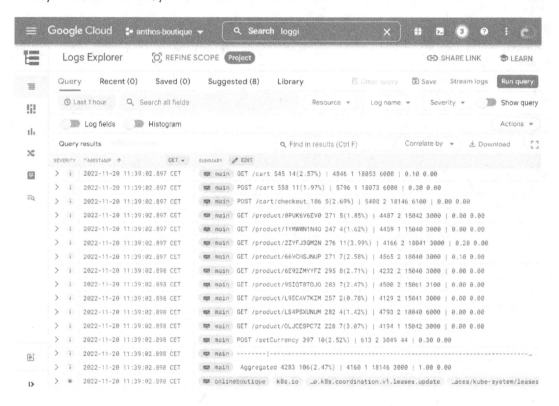

Figure 3.11 – The Cloud Logging main screen

Lucky for us, by default all platform logs are automatically ingested and stored with no previous setup required. GKE workload logs are also automatically captured too and, if we deployed Ops Agent in our VMs, their workload logs will also be automatically captured. Of course, we can use *exclude* and *include* filters to discard messages that we don't want or to choose exactly what we want to see in our logs.

Event filtering can be done in two different ways:

- Using *include* filters, where you specify patterns just for the events that you want to keep, while the rest are discarded. While this can seem an easy way for keeping only the events that you are interested in, it can also be a trap since you may be leaving out very important events or you can miss new important patterns that start being written to Cloud Logging after an update.

- Using *exclude* filters, where you define patterns for those messages that you don't want to see, is the method that I always recommend. While building the list of unwanted patterns can take a while and will require periodical reviews, any rare or new messages will be captured, providing better chances of identifying critical issues.

Now that Cloud Logging is ready, let's discuss some of the best practices for sending the most useful logging events from our application or service.

Best practices for logging

Implementing proper logging in our applications and services can make a difference when monitoring and troubleshooting our code since it can help us find the root cause for an issue much faster by identifying which part of our code failed and what the symptoms were.

First of all, it's important to adjust the **verbosity level**, that is, what kind of events we send and how often we do it. This is because logging too many events or doing it too frequently can also have a negative performance impact on our application, fill our logs with noise, and increase the cost if the volume is huge, especially when the application is working as expected, due to the unnecessary use of resources to write events that nobody will probably ever care to take a look at.

For this reason, having the option to temporarily increase the verbosity of the logger without having to make changes in the code or even recompile it can be very beneficial once an issue has been detected. Using either a `debug enabled` flag, which you can set to `true` to increase the verbosity, or a variable that allows customizing the *minimum* level of logging that will be sent to Cloud Logging (i.e., if the logging level is set to `warning`, then only events of severity `warning` and `critical` will be logged) can help us troubleshoot in both a faster and more efficient way.

In the world of microservices, changing the verbosity level should be a trivial change that can be applied just to a specific microservice by changing a configuration parameter, and be deployed and live in a matter of minutes, when we should start seeing more events almost immediately in our Cloud Logging console.

Another key topic is to make sure that each logging event sent contains all the information needed to make it useful. A good line of logging should include the following:

- A complete timestamp, including time zone information, especially for architectures distributed across data centers in different parts of the world

- The name of the microservice or component that sent the event

- The ID of the process, service, or instance the event refers to, which allows tracing back by looking at all events sent from that specific component

- A descriptive name of the part of the code that is running, such as a module, function, or action name

- Information about either the input or output variables, or the intermediate or final status of the operations being run by the portion of code the event is referring to

A generic line such as the following provides very few details about an error and is virtually useless:

```
2023-03-24 Data Reader error
```

But working just a bit to properly format the line and to make sure that all required information is present can make it much more useful for troubleshooting:

```
2023-03-24 23:09:14 CET [cache-loader] PID: 12217, File:
12217-cache.txt (Load Cache) Error reading file: Cache file
does not exist
```

This second example tells us when the issue happened (and in which time zone), what file was affected (`12217-cache.txt`), in which component (`cache-loader`), which operation was running at that time (`Load Cache`), and what was exactly the issue detected (`Cache file does not exist`). This information can help us understand where the error happened and start tracing back in the code to try to identify its root cause and either mitigate it or fix it.

Generating this kind of log event is very easy if you write a small function with parameters that generates the text for each event and combine it with a global on/off debug or a minimum severity flag to define in which specific cases an event will be sent. We will see a practical example of this implementation in the next chapter.

Once our code is ready and our logging configuration is complete, it's time to start testing it. Since the testing process is more complex than the rest of the steps covered in this chapter, I moved it to an appendix at the end of this chapter.

Monitoring the execution of your code

Monitoring is defined in this Google Cloud documentation page (`https://cloud.google.com/monitoring`) as the process of gaining visibility into the performance, availability, and health of your applications and infrastructure.

Introducing observability

In my professional career, I have met quite a few developers who thought that their work was done when the last system tests were passed, only expecting issues to be reported and fixed as part of the maintenance phase. They couldn't be more wrong, because once you test and deploy your code, you need to put in place a proper monitoring system to ensure that you can answer the following three questions at any time:

- Is your code actually running? If the process crashes and you never noticed, then you have a serious problem for sure.

- Does the performance of my code meet the requirements? Code running doesn't mean code running efficiently. You should define and measure **Service-Level Indicators** (**SLIs**) (what to measure) and establish **Service-Level Objectives** (**SLOs**) (what those measures should look like) so you can detect and quickly take action when an issue that is affecting the performance of your service is detected.

- Is your code really providing the expected services? Since a running process doesn't guarantee that the service it should be providing is actually working, there should be tests in place that periodically check the end-to-end availability of each service.

Observability, also known as visibility, is a key topic in the area of monitoring because we aim to obtain the information required to be able to answer these three questions at any moment in time. This connects with the principles of **Site Reliability Engineering** (**SRE**), where being able to detect potential issues and fix them before our users are affected can make our applications and services better as we make changes to improve their availability and performance, thus improving their overall reliability.

Gathering information about your services

In order to achieve observability, we need to understand our code and the services it provides and identify the key metrics that can be used to detect performance bottlenecks, integration issues, or global availability problems.

For example, if our service analyzes pictures of documents, extracts the text using **Optical Character Recognition** (**OCR**), and stores the result in a text file for each document, we could use the number of documents pending to be analyzed as a key performance metric, together with the ratio of errors returned by the Cloud Vision API (used to implement OCR) every 1,000 calls, or the number of output text files pending to be written to storage.

We could also measure memory and CPU utilization, and how many microservice instances are running every 5 minutes. This, together with the number of requests analyzed every hour, can provide us with a basic level of visibility of our service.

In an example like this, there would be different metrics involved:

- Internal metrics, such as the size of the document queue, the number of text files pending to be written to storage, or the number of hourly errors returned by the API, which can be logged periodically or obtained by parsing the logs externally.

- Service metrics, such as requests handled hourly, can be obtained by parsing log files externally.

- Infrastructure metrics, such as resource utilization, number of processes running, or number of instances of a microservice. These should be obtained by either using an agent running on the system or reading metrics provided by a hypervisor or a management API.

As you can see from the list, we need a few components to build a decent monitoring system:

- Periodical internal statistics and metrics measured and exported by our own code regarding health and performance

- System statistics and metrics provided by an agent running on the host OS

- Infrastructure statistics and metrics provided by a management API or a hypervisor

All this data, stored in a centralized database, will allow us to analyze trends, define a threshold, and set up alerts when the value of a metric is outside the usual range.

Google Cloud provides a set of tools to make this process easier:

- **Cloud Monitoring** (`https://cloud.google.com/monitoring/`) is also part of the Google Cloud operations suite and lets you access over 1,500 cloud monitoring metrics from Google Cloud and Amazon Web Services using its API (`https://cloud.google.com/monitoring/api/v3`), a list that can be extended by adding your own custom metrics. You can also use it to generate alerts when the value of one or more of your key metrics exceeds a defined threshold.

- Cloud Monitoring also provides predefined and custom **monitoring dashboards** (`https://cloud.google.com/monitoring/dashboards`) that let you combine data from different sources in the same context and create a visualization of the health of a service or application. These charts can be useful for operations and troubleshooting teams (including SREs) in order to quickly identify issues and their root cause, so the **Mean Time to Recovery (MTTR)** of the service can be kept to a minimum.

- **Ops Agent** (`https://cloud.google.com/monitoring/agent/ops-agent`) is the software that can obtain log events and metrics from Compute Engine instances, thus improving our global visibility of Linux and Windows VMs.

- If you have a big architecture or want a more powerful monitoring system, **Google Cloud Managed Service for Prometheus** (`https://cloud.google.com/stackdriver/docs/managed-prometheus`) provides a fully managed multi-cloud solution that is compatible with open source monitoring and dashboarding solutions, including Prometheus (`https://prometheus.io/`) and Grafana (`https://grafana.com/grafana/`). The benefits of using Managed Service for Prometheus are that you can monitor the whole infrastructure with a unified solution, centralize all the data gathered from the different sources, and use powerful queries to obtain and visualize the exact information and insights that help you achieve observability, and then set up alerts to be the first one to know when there is an issue. This solution works for both Kubernetes and VM workloads and has 24 months of **data retention**.

As you can see, monitoring our applications and services is very useful, but can also become quite complicated, especially as the number of microservices and components in use starts to grow exponentially.

While there are numerous third-party solutions specialized in cloud monitoring, Google Cloud offers a good set of tools that can help us set up our monitoring architecture and, ultimately, use it to achieve observability of our workloads and ensure that both availability and performance are within reasonable limits, and get alerted when something wrong happens, so that we can troubleshoot our code and find out what's going on.

Troubleshooting by debugging, tracing, and profiling your code

Our code can have issues of different types. For example, it may just not run as expected due to a bug, or it may be running apparently fine, but have some operations lasting much longer than initially expected. In situations like these, we will need to dig deep into our code, follow the execution flow step by step, and ultimately identify what is wrong and put together a fix.

While this can be easy to do locally, it may become quite complicated in cloud environments, especially with technologies such as clusters, where observability is limited, and you usually need to resort to remote debugging techniques if you want to have some visibility. Fortunately, Google Cloud provides a set of tools that makes these tasks much easier, helping us avoid tedious tasks such as port forwarding, even if our workload is running on a different cloud provider.

With Cloud Code, for example, you can debug remote containers in the same way as if they were running locally. Isn't that great? Let's describe how debugging works, and we'll see some real examples in the next chapter.

As we already discussed, we can either use Cloud Shell Editor or our favorite supported IDE to enjoy the features of Cloud Code, one of them being to help us debug our code. In the case of container debugging, all we need to do is to have a local copy of the code running on the container, run the container in debug mode, and attach our remote debugger to the corresponding **Artifact Registry** or pod, including support for Google Cloud, Amazon AWS, and Microsoft Azure registries, and we will be able to see the real-time execution logs, inspect variables, and set breakpoints as if we were working with a local container.

Cloud Code will help us through all the steps, including the configuration, if it wasn't already completed, of both the **Skaffold** (`https://cloud.google.com/skaffold`) and the `cloudcode.kubernetes` launch configuration.

Debugging our code can be very helpful because we can check the value of all related variables around the portion of code where an error is reported, and start tracing back in our code until we find the line where one or more variables get a wrong value. Since we have access to the source code locally, we can take study the code, compare it with the original requirements, and, fingers crossed, ultimately identify what is wrong. We can even change the value of the affected variable in real time, after setting a breakpoint on the offending line, and verify that it works fine before starting to write a patch to fix the issue.

This is what a debugging session looks like using VS Code:

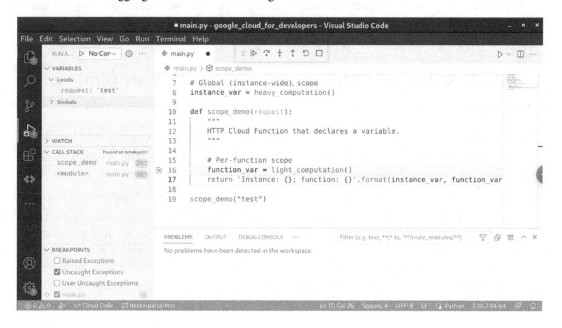

Figure 3.12 – Debugging in VS Code

There is even a `watch` mode that detects any changes that you perform locally on the code and automatically redeploys a new version of the container and reconnects to it, making it even faster to verify that the changes in the code actually fix the problem by testing it on a live cluster. This mode can, of course, be disabled if you are not comfortable with this process and prefer to decide when changes should be applied remotely.

If we detect a performance issue, sometimes, we may be able to apply a quick fix by reordering our code or optimizing loops, but there will be cases where we will need a deeper understanding of which parts of our code take longer to run, so we can focus on improving their performance. This is when **Cloud Trace** and **Cloud Profiler** come to the rescue!

As we can read on its documentation page, Cloud Trace (`https://cloud.google.com/trace`) is a distributed tracing system that collects latency data from our applications and displays it in the

Google Cloud console. We can track how requests propagate through our application and receive detailed near real-time performance insights.

Cloud Trace automatically analyzes all of our application's traces to generate in-depth latency reports to surface performance degradations and can capture traces from all of our VMs, containers, or App Engine projects.

Indeed, all Cloud Run, Cloud Functions, and App Engine standard applications are automatically traced and it's very easy to enable tracing for applications running elsewhere.

Cloud Profiler (`https://cloud.google.com/profiler/docs`) is a statistical, low-overhead profiler that continuously gathers CPU usage and memory-allocation information from our production applications. It attributes that information to the application's source code, helping us identify the parts of the application consuming the most resources.

The main benefit of combining these two tools is that they provide performance observability, helping us understand which parts of our architecture in general and our code in particular are degrading the global performance or eating too many resources. This information is especially valuable in a cloud environment because we can easily differentiate which performance issues are being caused by infrastructure components, and take strategic decisions to mitigate them, and which ones are caused by our code, and in that case, work on replacing specific libraries or improving the performance of specific portions of code to make more reasonable use of the allocated resources.

A well-performing application will not only provide a faster service but also help our organization save money by reducing the number of resources required, being able to serve more users with the same resources it used before being optimized.

We will see practical examples of debugging, tracing, and profiling for each of the options to run our code on Google Cloud in the next chapter.

Appendix – testing your code on Google Cloud

I added this section as an appendix because testing is a harder concept that requires more thought and customization, so I consider this as an extra effort that will be worth your while.

When we speak about testing, there are a few basic concepts that we should take into account, as this Google Cloud documentation page mentions: `https://cloud.google.com/functions/docs/testing/test-basics`.

We should ensure that our code works properly from different perspectives. Let's introduce the three types of tests.

Types of tests

An important concept to keep in mind is that a portion of code, even if it implements a full service, needs to be tested from different points of view before we can say that it fully works as expected. And for this to be possible, we need to put together a list of tests that, when passed, will confirm that our code is functional in three different scenarios: on its own, as part of a bigger workflow, and as part of a whole system.

First, **unit tests** help us test code on its own, just taking into account its expected functionality, the edge cases, and assumptions that we considered at the time when the code was written. We can verify this by providing inputs to the code and comparing the results obtained against those expected.

Unit tests are defined by developers to ensure that changes in code don't break basic functionalities. In this phase, we do not integrate our code with other components, but we rather replace them with mocks and emulators.

If we think of a car, a unit test would involve testing a tire or the brake pedal separately from the rest of the car, even those parts that usually interact with them.

Then, **integration tests** help us verify that our code integrates correctly with other services and components, which means that mocking must be kept to a minimum and we need to build end-to-end tests involving any other components and cloud services that are used to provide its functionality.

Integration tests help us validate code for a microservice as a part of a bigger workflow, service, or operation, and verify that it communicates properly with the rest of the microservices and components involved. The integration with any external platform services and components used by the code to be tested is also validated by identifying end-to-end operations where parameters are provided and a final response is provided after following a specific workflow, making it possible to detect whether either an issue in our code or an issue or change in another component made the whole operation fail or provide a response different from the one expected.

In the example of a car, an integration test could be used to validate the integration of all the components of the braking system, including the expected response of a car when we hit the brake pedal until the expected outcome happens (wheels no longer move and the car stops).

Finally, **system tests** validate the whole functionality of an application or service. We can understand system tests as a series of integration tests run together, and sometimes involving connections between the operations validated using integration tests. These tests help us validate that the whole system, including all its services, meets the defined business requirements.

In the case of Cloud Code, some components may be reused in a different part of the architecture, and system tests would validate each of these uses, to make sure that all of these components are working as expected.

In the example of a car, we should run tests for all elements and features of the car: lights, speed and power, security, safety, and many others. As you can see, as we move from unit to integration, and from integration to system tests the complexity increases, as does the number of components involved and the time required to complete them.

Recommendations and best practices for testing your code

As you can see after reading the previous section, testing can become quite complicated as our architectures begin to grow, and automating the testing process can help us save a lot of time and eliminate human errors, helping us detect issues faster.

But before we are able to actually implement this automation for tests, we should take a few tips and best practices into account.

When possible, **separate code and configuration**. The main reason for this suggestion is that sometimes issues can be fixed just by making a change in the value of one or more configuration parameters, and it can be much more time-consuming if you need to rebuild the code than if you can just make a change in a JSON or XML file, or an environment variable and restart the service (or even make the service reload its configuration without having to restart it).

Since some of the configuration parameters may include passwords and other information that should be properly protected, you can use Secret Manager (`https://cloud.google.com/secret-manager`) to securely store sensitive information and retrieve it.

Another good practice is to use **hermetic builds** (`https://sre.google/sre-book/release-engineering/#hermetic-builds-nqslhnid`), meaning that your builds are consistent and repeatable, that is, not dependent on the version of the tools running on the build machine, but depending only on known versions of compilers and libraries. This will provide a controlled environment, where the same build process for the same code version in two different build machines will provide identical results. This type of environment makes testing much easier because we avoid failures caused by external components, often not under our control, that may complicate troubleshooting a lot.

It's also strongly recommended to implement a proper **retry strategy** (`https://cloud.google.com/storage/docs/retry-strategy`) in your code. An API call failing once doesn't mean that our code doesn't work and, as we discussed earlier in the book, all code in general, but code to be run on the cloud in particular, should be ready for failures and configured to retry any operation that fails, whether it's an **idempotent operation** or not. Concepts such as **exponential backoff**, where the waiting time between consecutive retries grows exponentially, can help our code elegantly handle temporary failures while eventually passing all tests and being resilient when deployed in production.

Finally, all the aforementioned suggestions will make it easier for you to implement **Continuous Integration and Deployment** (**CI/CD**) pipelines to complete both your unit tests and your integration tests on Google Cloud, implementing what is known as **continuous testing**. **Cloud Build** (`https://cloud.google.com/build`) can run your tests on an ongoing basis, helping you to ensure that your code keeps on working as expected and that all dependencies are up to date.

We will discuss practices such as exponential backoff and concepts such as continuous testing, as well as go through practical examples, in the next chapters of the book, where we will look into the different options to test, run, and troubleshoot code on Google Cloud.

Summary

In this chapter, we covered the tools that Google Cloud provides for the different phases of software development, and discussed their features and how they can be of help when writing, running, testing, and debugging code to be run on the cloud.

First, we introduced Cloud Shell and Cloud Shell Editor and mentioned how Cloud Code can help us integrate code writing and testing for different Google Cloud products. We also covered alternative IDEs, such as VS Code, that Cloud Code is also compatible with, and then talked about Cloud Logging and the importance of setting up proper logging in our applications.

Then, we talked about the need for proper monitoring to achieve observability and closed the chapter by explaining the available tools for troubleshooting issues, including debugging, tracing, and profiling our code in order to fix availability issues or to improve its performance. Finally, we enumerated the different types of tests and provided some tips on how to set up a good test environment.

This is the last chapter with a more theoretical focus. Starting with the next one, we will deep dive into the different options for running code on Google Cloud, with a lot of practical examples where we will put into action all of the topics that we covered in these first three chapters.

Further reading

To learn more about the topics covered in this chapter, please visit the following links:

- *Tips and Tricks for using Google Cloud Shell as a Cloud IDE* (`https://dev.to/ndsn/tips-and-tricks-for-using-google-cloud-shell-as-a-cloud-ide-4cek`)

- *Building idempotent functions* (`https://cloud.google.com/blog/products/serverless/cloud-functions-pro-tips-building-idempotent-functions`)

- *Concepts in service monitoring* (`https://cloud.google.com/stackdriver/docs/solutions/slo-monitoring`)

- *Setting SLOs: a step-by-step guide* (`https://cloud.google.com/blog/products/management-tools/practical-guide-to-setting-slos`)

- *Observability in Google Cloud* (`https://services.google.com/fh/files/misc/observability_in_google_cloud_one_pager.pdf`)

Part 2: Basic Google Cloud Services for Developers

One of the benefits of running our code on Google Cloud is that we no longer need to use servers. This part of the book will cover three different options for running your code using a serverless approach.

We will then deep-dive into the concept of containers and how they can be used to abstract even more of our code from the underlying infrastructure.

Finally, we will explore how Anthos can make it easy to design hybrid and multi-cloud architectures, while at the same time, it provides global observability and the ability to move any workload from one provider to another, including private on-premises clusters and multiple public cloud providers, so that you can make the most out of each option.

This part contains the following chapters:

- *Chapter 4, Running Serverless Code on Google Cloud – Part 1*
- *Chapter 5, Running Serverless Code on Google Cloud – Part 2*
- *Chapter 6, Running Containerized Applications with Google Kubernetes Engine*
- *Chapter 7, Managing the Hybrid Cloud with Anthos*

Running Serverless Code on Google Cloud – Part 1

After three chapters without having written a single line of code, you will probably be looking forward to some hands-on action happening as soon as possible. As you will see, it was worth the wait, since I introduced a lot of concepts that we will be using in this and the following chapters.

This chapter will cover two of the serverless options available for running your code on Google Cloud, while the next will cover the third. You will learn what serverless means and then I'll introduce each of the serverless options, together with an example that we will make run in each option so that we can compare the implementation, as well as tips for running your code on Google Cloud using either Cloud Functions or App Engine in this chapter, or Cloud Run in the next one.

Finally, we will discuss their similarities and differences and when you should choose each.

We'll cover the following main topics in this chapter:

- Introducing serverless architectures
- Using Cloud Functions to run your code
- Using App Engine to run your code

Let's get started!

Technical requirements

If you want to complete the exercises included in this chapter, all you will need is access to the Google Cloud console, a Google Cloud project with either billing enabled or some available free credits, and the code files for this chapter, which are available in the code repository for this book: `https://github.com/PacktPublishing/Google-Cloud-for-Developers`.

Introducing serverless architectures

For decades, the infrastructure required to run code included an **Operating System (OS)** running on top of dedicated hardware, leading to a tremendous waste of computing resources.

While virtualization started in the late 1960s for mainframes, it wasn't until the early 2000s that it became generally available and users could finally share resources, which started to simplify the original scenario. Virtualization created multiple logical servers on top of a shared pool of computing power, allowing for allocated resources to be better adjusted, and providing services to more users with the same or less hardware.

The use of containers, whose predecessors we've been using since the 1970s, exploded in popularity when Docker emerged in the early 2010s. Using containers reduces the contents of a deployment package to just the OS libraries and the dependencies that our code requires, making packaged applications much smaller and also portable, with a higher level of abstraction because a common host OS is shared by all the applications or services running in containers.

The use of these and similar technologies led to the appearance of different levels of abstraction of hardware and OSs, eliminating the complexity of setting up and maintaining the underlying infrastructure, among which I would like to highlight the following:

- **Function as a Service (FaaS)**: Runs a function of code in an environment that scales according to the traffic

- **Platform as a Service (PaaS)**: Runs a frontend or backend application and adjusts the resources according to the traffic and the load

- **Container as a Service (CaaS)**: Runs a container and adjusts the number of instances depending on the load

Google Cloud offers three different products to run our code using serverless architectures, where we care about packaging and deploying our code and Google Cloud takes care of all the rest. Patching, maintenance windows, and updates are no longer taking most of our time, and we can now dedicate our efforts to writing even better code, focusing on innovation at its best. Doesn't this ring a bell? I hope it does...

Let's discuss each of the three options available to run your code using a Google Cloud serverless product: Cloud Functions, App Engine, and Cloud Run.

Using Cloud Functions to run your code

Using **Cloud Functions** (`https://cloud.google.com/functions`) is one of the simplest ways to run your code on Google Cloud, using the benefits of a FaaS platform. Let's explain what it is and how it works.

Introducing Cloud Functions

The concept of Cloud Functions, and the reason for its name, is that you can create a code function in your favorite programming language and use a trigger to execute it on demand. Personally, I love the idea of packing my code, specifying the dependencies in a text file, deploying it, and… voilà, it is ready to run in a matter of minutes with no hassle.

The choice of programming language is quite extensive, including Node.js, Python, Go, Java, .NET, Ruby, and PHP at the time of writing. You can see the full list of supported languages and versions here: `https://cloud.google.com/functions/docs/concepts/execution-environment#runtimes`.

There are different triggers available to get your code to run:

- **HTTP/HTTPS**: Make our function react to HTTP requests.

- **Cloud Storage**: Run our function when a file is uploaded or updated in **Google Cloud Storage (GCS)**.

- **Pub/Sub**: React to a new message being received.

- **Cloud Scheduler**: Integrate our functions with a modern version of cron.

- **Cloud Tasks**: A great option for repetitive actions with high volumes that supports parallel calls to Cloud Functions to process noticeably big queues with thousands or even millions of invocations. Service requests are sent to Cloud Tasks, which handles multiple threads with automated retries and configurable exponential back-off.

You can read the full up-to-date list of triggers in this section of the official Google Cloud documentation: `https://cloud.google.com/functions/docs/calling`.

As you can see, the combination of the most used programming languages and a complete set of triggers makes Cloud Functions a valid solution for many frontend and backend scenarios.

Let's list some of the most common combinations, so you can better understand what we can do with them:

- **Run a basic web service**: For instance, we can get some parameters from the HTTP request, run an operation with them, and return a JSON file with the results. Or, we can even implement a small web server and return HTML.

- **Process files as soon as they are uploaded or modified in GCS**: For example, generate a thumbnail automatically for each new image that is uploaded to a GCS bucket. We can also use triggers for files being updated, archived, or deleted.

- **Handle a queue of requests**: For instance, we can translate each of the comments received in any other language to English during the last hour. We can send hundreds of asynchronous requests using **Pub/Sub** messages from another microservice and each of them will be shortly processed by a different instance of a cloud function. There are other more efficient architectures for this kind of situation, as we will discuss a bit later in this chapter.

- **Run a workload periodically**: For example, train a machine learning model weekly, every Monday morning. Using **Cloud Scheduler** to schedule the trigger, we can either use HTTP or send a Pub/Sub message every Monday at 6 A.M. to start the cloud function and use the Vertex AI API to re-train our model.

> **Note**
>
> There are currently two generations of Cloud Functions coexisting at the time of writing this chapter. I will focus code and comments on the second generation since it is much more powerful and it's constantly receiving new features, while the first generation is more limited and will probably not be updated any longer. You can read more about the differences between these two generations in this section of the documentation site: `https://cloud.google.com/functions/docs/concepts/version-comparison`.

Now that we have a clearer picture of what Cloud Functions is, let's discuss how it works and what the key technical aspects are that you need to know so that you can decide whether it is the best fit for your development needs.

The following paragraph from the documentation page perfectly describes the inner workings of Cloud Functions and can lead us to a few interesting discussions (`https://cloud.google.com/functions/docs/building`):

When you deploy your function's source code to Cloud Functions, that source is stored in a Cloud Storage bucket. Cloud Build then automatically builds your code into a container image and pushes that image to the Artifact Registry. Cloud Functions accesses this image when it needs to run the container to execute your function.

So, in summary, we write a function in our favorite programming language, it gets automatically containerized and associated with a trigger on deployment, and it ultimately becomes a cloud function that runs on demand when the configured trigger is detected.

> **Note**
>
> Cloud Functions is billed based on how many times it is executed, apart from the costs of any other Google Cloud products that we may invoke from our code or use as part of the architecture used to invoke Cloud Functions.

Since its code is run on demand, if we don't use our Cloud Functions instance for a while, it will be shut down to save resources (and cost) unless we specify to keep a specific number of instances always running. This is called **zero-instance scaling** and means that when we use it again, the first execution may take a bit longer because the container needs to be cold-started again. You should take this into account, especially if you will be using Cloud Functions in real-time scenarios, where you should either define a minimum number of instances to always be running or use some less elegant alternatives. An example is periodically invoking your Cloud Functions instance to *awaken* or keep it running, for example, using a parameter that just wakes up the Cloud Functions instance.

When we deploy a cloud function, there is a set of parameters that we can configure and that will be key to its performance:

- The region where the container will run, which can affect the latency of our users. Of course, we can replicate cloud functions in different regions.

- The programming language chosen to write the code of the function.

- The amount of memory to be allocated (and, as we will see later, this also decides the associated amount of vCPU). The defaults are 256 MiB and .167 vCPU, which is a sixth part of a 2.4 GHz CPU.

- The timeout, in seconds, before each execution is automatically killed. The default is 60 seconds.

- The maximum number of instances running in parallel, so we can limit the maximum scaling.

When our cloud function needs to handle multiple requests, there are at least two different ways of doing it:

- The first one is quite simple: we just let each instance of our cloud function handle exactly one request. This makes the design much simpler but may increase the costs since we are billed by the number of executions and, depending on the scenario, with this approach, there may be millions.

- The second choice is to take advantage of the maximum execution time, which is 540 seconds for first- and second-generation cloud functions, except for second-generation HTTP-triggered functions, where they can run for up to 60 minutes and try to run as many complete operations as possible during that time. This requires a bit more complexity because we should treat each operation as a transaction and only remove it from the queue, marking it as complete, once the whole transaction has been successfully executed. But on the other side, it can speed up the execution time required to process the whole queue, and make your code more cost-effective.

Which of the two approaches should you use? Well, it will depend on your use case and the number of expected runs every month. You will probably need to make some calculations and compare the simple approach against the more complex one and decide whether it's worth the extra effort.

Before getting to the example, let's discuss the concept of a service account, which we will use constantly in all serverless products.

Running code using service accounts

When we log in to the Google Cloud console, we have a set of roles and permissions granted to our account by the admin using **Identity and Access Management (IAM)**, which defines what we can and cannot do while we interact with the Google Cloud console.

When we need to run code or start a virtual machine, we will need to associate it to an account too, for the same reason. If we used our own account, it would be inheriting our roles and permissions, and this could be a source of trouble for different reasons:

- We would be breaking the principle of least security, running code with more permissions than required, which would be an important security risk

- If the user leaves the organization, we will need to transfer the ownership of all the code running from the original account, probably update the roles and permissions of the receiver, and finally redeploy all assets, which would be really inconvenient

Service accounts were created to solve both problems. They are non-interactive accounts that we use to run services and deployments and that can have roles and permissions assigned to them too.

For example, if we want to run a cloud function, we can configure it to run using a specific service account. In that case, our code will run authenticated as this service account, and we can use IAM to grant additional roles that will allow our cloud function to access any additional resources required to complete its tasks. In this case, the service account would be the identity of the application, and the associated roles would be used to control which resources the application can access.

A default service account is automatically created with a limited set of permissions in each Google Cloud project, and it is assigned by default to all deployments. However, those permissions will often not be enough, and in that case, we can create our own service account, use IAM to grant it the minimum set of permissions required by our code to complete its tasks, and redeploy the cloud function using this new service account. This same approach can be used with any other serverless product.

There are some differences between service accounts and user accounts that you should be aware of:

- While user accounts must have a password set (and should have a second authentication factor enabled for a higher security level), service accounts don't use passwords and cannot be used to log in using a browser or cookies.

  Instead, we can generate public and private RSA key pairs to authenticate service accounts and sign data.

- Users or service accounts can impersonate another service account if they are granted specific permission to do so, which can significantly simplify permission and account management.

- Service accounts are totally separated from Google Workspace domains, which means that globally shared assets in a domain will not be shared with service accounts and, vice versa, assets created by a service account will not be created in your Workspace domain.

You can read more about service accounts on this page of the official documentation site: `https://cloud.google.com/iam/docs/service-accounts`

Now, finally, it's time to start writing our first cloud function.

Writing, deploying, and running a cloud function

I will be running away from "Hello world!" examples in this book and will try to include instead useful examples that can run using the Free Tier.

> **Note**
>
> Google Cloud provides Free Trial and Free Tier (`https://cloud.google.com/free`) options for some of its services, free of charge every month. Knowing the details about these can be very interesting, since you can use them to run the examples provided in this book, test new services, reduce your costs, or host your resume for free!

In the case of Cloud Functions, the following services are included in the monthly Free Tier:

- 2 million invocations per month (including both background and HTTP invocations)
- 400,000 GB-seconds and 200,000 GHz-seconds of compute time
- 5 GB network egress per month

Now, let's make use of our Free Tier to learn how Cloud Function works.

The example that I mentioned earlier was a website that uses parameters to send a response, but I also mentioned that it could be used as a web server. Let's combine both ideas and build a cloud function to host our resume so that we can share it with anyone interested and provide a certain degree of personalization for each recipient.

You can use this example to highlight your knowledge of Google Cloud. What would be a better way to do so than using a cloud function to share your resume with interviewers and hiring managers? We will later implement this example with the other serverless products too, so we can see the similarities and differences between platforms.

If you remember the concept, a cloud function has an entry point, a function, which is run when it is triggered by an event. Taking this information into account, we can organize our code so that it can be compatible not only with the deployment of a Cloud Function but also with unit tests.

For this, we can include a main function that will never be executed on the cloud but will allow testing the code using the command line. We can also divide our code into functions that implement a separate part of the whole process so that we can later test one or more of those parts whenever we need to.

Let me show you how to do this using our example about resumes, which includes the following actions:

1.   Load an existing HTML resume template from GCS.
2.   Check passed parameters and build a customized header message.
3.   Replace the placeholder in the resume template with a customized message.
4.   Return the full resume to the requestor.

So, if we split the code into four functions, each taking care of one of these topics, we can have different functionalities that we can test separately. We can even include unit tests as part of the cloud function in our repository. They will be used for local testing purposes. A `main` function in the Python file will enable us to run the full code, or just a part of it, using the command line, which can speed up testing even more.

I will now show you how I got to the files included in the repository for this chapter. You may refer to the `Cloud Functions` directory to see the whole code, which I will reference in the next part of this section.

First, let's create an HTML file for the resume. I will name it `english.html` and include a few lines and an externally hosted picture. This is just a simple example. I'm sure you can design a much better resume, in terms of both content and design. I have included a placeholder tagged as `##RESUME_ HEAD##` at the top of the HTML file that we will replace with a proper title right before returning the full resume to the requestor.

The HTML file looks like this when loaded on a browser. Notice how the first line has been personalized for a fictional interviewer:

*(Specially prepared for **John Smith** from **StarTalent**)*

Jane Doe
211, Short St
New Jersey 07070
USA

Skills

- Expert developer on Google Cloud.
- Certified Red Hat System Administrator.
- Team Player and good Leadership skills.

Education

- *MBA at BizzNezz School.*
- *Bachelor of Science at Kalsh University.*

Professional Experience

- ***2021 - Today:*** *Cloud Admin at Storming Systems.*
- ***2018 - 2021:*** *IT Specialist at Loyal Bank.*
- ***2017 - 2018:*** *Trainee at Computer Land.*

Figure 4.1 – A preview of the resume in a web browser

Our cloud function will read the HTML template from GCS, customize the header depending on the parameters received, and return the final HTML to the caller, acting as a web server. This is a portion of the code that we should use for that purpose inside main.py:

```
def return_resume(template, name, company):
    resume_html = load_resume(template)
    resume_header = build_resume_header(name, company)
    resume_html = replace_resume_header(resume_html,
                                        resume_header)

    return resume_html
```

We will now code three more functions to load the resume, build the customized header, and replace it in the raw HTML:

```python
# Imports the Google Cloud client library
from google.cloud import storage

# Name of the bucket storing the template files
BUCKET_NAME = "resume_xew878w6e"

def load_resume(template):
  # Instantiate a Cloud Storage client
  storage_client = storage.Client()

  # Open the bucket
  bucket = storage_client.bucket(BUCKET_NAME)

  # And get to the blob containing our HTML template
  blob = bucket.blob(template)

  # Open the blob and return its contents
  with blob.open("r") as resume_file:
    return(resume_file.read())

def build_resume_header(name, company):
  custom_header = ""
  if name or company:
    custom_header = "(Specially prepared for "
    if name:
      custom_header = custom_header + "<strong>" + name +
                      "</strong>"
    if company:
      if not name:
        custom_header = custom_header + "<strong>" + company
+                       "</strong>"
      else:
```

```
            custom_header = custom_header + " from <strong>" +
                            company + "</strong>"
        custom_header = custom_header + ")"
    return custom_header


def replace_resume_header(resume_html, header_text):
    return resume_html.replace("##RESUME_HEAD##", header_text)
```

Now, all we need is to add the function that will be triggered by each HTTP request, together with a main function. I have written the trigger function as a wrapper so we can also run unit tests for each functionality, and added the main function so we can test the code just by running it from the command line:

```python
import functions_framework
DEFAULT_TEMPLATE = "english.html"
@functions_framework.http
def return_resume_trigger(request):
    template = request.args.get('template', DEFAULT_TEMPLATE)
    name = request.args.get('name', None)
    company = request.args.get('company', None)
    resume_html = return_resume(template, name, company)


def main():
    template = "english.html"
    name = "John Smith"
    company = "StarTalent"
    resume_html = return_resume(template, name, company)
    print(resume_html)


if __name__ == "__main__":
    main()
```

Using this code, we can run quick checks from the command line and also write and run unit tests, checking for the expected output of each function separately or all of them in sequence, in order to verify that all the functionality is intact after making changes to our code.

Once all tests are successfully passed, our cloud function should be ready for testing.

Testing a cloud function

As we will see shortly, deploying a cloud function can take a few minutes. While it's a straightforward process, if you want to follow a typical development cycle, first testing your code, then identifying issues, then getting them fixed in your code and re-iterating, it can become quite frustrating and inefficient because you will be spending more time staring at the screen waiting for the deployment to complete rather than coding or testing your code.

To make this process easier, Google Cloud provides a **Cloud Functions emulator**, which allows us to set up a local server that will simulate what the actual product does and enable fast testing by directly using our local code to serve requests. This way we can run tests and just deploy the final version once all of them pass locally. This doesn't mean that we should bypass tests on the cloud but will just make the first iterations much faster.

Installing the emulator is extremely easy. Just use the following `pip` command:

```
pip install functions-framework
```

Once the installation process is completed, you can change to the directory where the main source code file for your cloud function is located and run the emulator using this command:

```
functions-framework --target=return_resume_trigger
```

This will start a local server on port 8080 (we can customize the port, of course) that will execute the `return_resume_trigger` function on each connection and will return the results to the caller.

We can invoke the cloud function, if it is triggered using HTTP, as was the case for our resume example, running the following command and using double quotes to enclose the URL:

```
curl "http://localhost:8080?template=english.
html&name=John+Smith&company=StarTalent"
```

If you need to trigger background cloud functions that use Pub/Sub or GCS events, you can read how to do it on the following documentation page: `https://cloud.google.com/functions/docs/running/calling`

Notice how I'm passing the values for both name and company by double quoting the full URL so that all the parameters are passed to the server. If you don't use double quotes, your server will only receive the first parameter because the ampersand will be interpreted by the shell as the end of the command and no user or company name will be printed.

You should now see the HTML content of your resume printed on your screen, which you can validate by loading the URL in a browser. If there are any issues or you see anything that you don't like, just make the appropriate changes and try again. Once you are happy with the resume, we will be ready to deploy the cloud function to Google Cloud.

> **Important note**
> Each time you make changes to your code, you will need to stop and restart the test server for the code to be refreshed. Having a script to do this could be quite convenient.

If you are using other Google Cloud services, such as Pub/Sub, my recommendation is to find out whether an emulator exists from the documentation website before using real services for your tests, especially if your code triggers a service hundreds or thousands of times, so you don't incur significant costs. For example, if your code makes use of Pub/Sub, you can read more about its emulator in this section of the documentation: `https://cloud.google.com/pubsub/docs/emulator`.

Deploying a cloud function

The last step before having a cloud function ready for use is deploying it and we will need a deployment package to do it.

A deployment package will include at least one file with the source code, or multiple ones if we are using modules. We can also structure our code in subdirectories, and it will also work as soon as the imports are properly specified and resolved in the code.

We should also include any external or added files that our cloud function is using, such as the HTML file for the resume in our previous example. Please notice that these files will not be directly exposed with a public URL but we will need to read them from our code, instead.

Finally, when using Python as the programming language for Cloud Functions, as I will do for the examples in this book, we should use a file named `requirements.txt` to specify which external libraries we are using, so they can be installed before our function is executed. If we miss any libraries, the code will just fail to run because the Python interpreter will not be able to resolve that dependency.

Since `pip` is used to install the updates, the requirements file should contain one line per package, each including the package name and, optionally, the requested version. If a version is not specified, the latest one available will be installed.

The following is an example of a `requirements.txt` file for a cloud function requiring the use of the latest version of the GCS library and version 2.20.0 of requests:

```
requests==2.20.0
google-cloud-storage
```

> **Note**
>
> If you are wondering whether and why you should be including the required version for all libraries in your requirements file, you can have two different scenarios: first, if you don't specify the versions, each deployment will always use the latest versions of all libraries. While this can seem to be beneficial for the sake of security or to get bugfixes and optimizations deployed as soon as they are available, it can also mean breaking changes in major version updates, including changes in the list of expected parameters and other scenarios that you will probably prefer to avoid.
>
> On the other hand, specifying the requested version for all libraries guarantees that the code will always be using the same version of the libraries, but will force you to periodically test the newest versions and update the requirements file to prevent the versions used becoming out of date and even no longer being available.
>
> In my experience, a version freeze will cause you fewer headaches because issues due to updated libraries can happen at any time and they may surprise you in the worst moment when you don't have time or resources to deal with them. Freezing the versions will allow you to decide how often and when you will perform the library updates. This is an added reason why having a nice set of automated tests can help you quickly verify whether, after updating your requirements file to use the latest versions of the libraries, your code still passes all the tests.

You can read more about specifying dependencies, including the list of packages preinstalled in Cloud Functions using Python, on the following documentation website: `https://cloud.google.com/functions/docs/writing/specifying-dependencies-python`.

Our example uses GCS to host the HTML templates, so the first thing that we need to do is to create a new bucket and upload the template file there. You can use the following address and then select your project using the drop-down menu at the top, if required: `https://console.cloud.google.com/storage/browser`. Once there, just click on the **Create** button, and choose a unique name for your bucket, such as `resume_xew878w6e`, which I used in my example. Bucket names are globally unique, so sometimes it can take a while until you find a name that is not in use. Write down the name because you will need to fill it in the Cloud Function's `main.py` file.

Once you confirm the name, you can select **Region** hosting, since the simpler option will work for our test, and leave all the other options at their default values. Click on the **Create** button and confirm the prevention of public access, since we will use the bucket to store files internally. Now, you can use the **Upload files** button to select and upload the `english.html` template file from the source directory.

Once the template has been uploaded, in order to deploy the cloud function, we will just need a directory containing the `main.py` and `requirements.txt` files. Edit `main.py` and replace the value of `BUCKET_NAME` at the top of the file with the name of the bucket that you just created. Save the file and now it's time to prepare our environment for the deployment.

First, we will need `gcloud`, the Google Cloud command-line utility, to be installed. It comes preinstalled in Cloud Shell but if you use any other environment where it is not installed yet, just run this command or a similar one compatible with your development environment:

```
sudo apt-get update && sudo apt-get install google-cloud-cli
```

If `gcloud` was already installed, you can search for available updates by running this command:

```
gcloud components update
```

Now, it's time to authenticate ourselves and set the default Google Cloud project for the gcloud utility by running the following two commands and following any instructions that they specify:

```
gcloud auth login
gcloud config set project <your_project_name>
```

If we have any extra files in our Cloud Function's directory, we can prevent them from being deployed to the cloud using a `.gcloudignore` file, so that unit tests, temporary files, and similar examples never get deployed. Each line of this file contains either a complete file or directory name or a pattern that will be checked against each filename before deciding whether it will be deployed or not.

This would be the sample content of a `.gcloudignore` file to filter out Git files:

```
.git
.gitignore
```

You can read more about this feature by running the following command:

```
gcloud topic gcloudignore
```

Check whether `gcloudignore` is enabled by running this other one:

```
gcloud config list
```

Finally, enable it, if it wasn't already, using this final one:

```
gcloud config set gcloudignore/enabled true
```

Now, we are ready to deploy the cloud function using the following multi-line deployment command, which includes default values for the most common parameters:

```
gcloud functions deploy resume-server \
--gen2 \
--runtime=python310 \
--region=us-central1 \
--memory=256MB \
```

```
--source=. \
--entry-point=return_resume_trigger \
--trigger-http \
--allow-unauthenticated
```

Please take a look at the command, so you can understand the different configuration options we are setting here:

- The function name will be `resume-server`
- It's a second-generation function
- It will be running from the `us-central1 region`
- It will be limited to 256 MiB and .167 vCPU
- The source code to be deployed will be in the same directory from where we are running the `deploy` command
- The function to be executed by the trigger will be `return_resume_trigger`
- This is an HTTP-triggered function
- We are allowing unauthenticated users to run this cloud function, that is, making it public and open for anyone to run it if they know the URL to use

All these options, and many more, can be customized using the different command-line parameters of the `build` command, as described in the corresponding documentation section: `https://cloud.google.com/sdk/gcloud/reference/functions/deploy`

The first time that we run a `deploy` command, we will be requested to enable the APIs for Cloud Functions, Cloud Build, Artifact Registry, and Cloud Run if they weren't already so that the deployment can be completed. Just answer `y` for each of the requests and the deployment will begin:

```
API [artifactregistry.googleapis.com] not enabled on project
[<your_project_name>].
Would you like to enable and retry (this will take a few
minutes)? (y/N)? y
```

Once you answer positively, you will see how Cloud Run is containerizing your cloud function and the console will keep you informed about each step until the trigger URL will be displayed.

We will be using the Free Tier to store our cloud functions without additional costs unless you already have many other cloud functions deployed in your project and exceed the free quota. An excerpt of the output should look like the following:

```
[...]

Preparing function...done.
```

```
OK Deploying function...
[...]Done.
You can view your function in the Cloud Console here:

[...]

  timeoutSeconds: 60
  uri: https://python-http-function-4slsbxpeoa-uc.a.run.app
```

As you can see in the preceding code, my external URL for the cloud function would be `https://python-http-function-4slsbxpeoa-uc.a.run.app`. But since it's unique for each project, you will get a different one for yours, and loading that URL in a browser will actually display our sample resume:

*(Specially prepared for **John Smith** from **StarTalent**)*

Jane Doe
211, Short St
New Jersey 07070
USA

Skills

- Expert developer on Google Cloud.
- Certified Red Hat System Administrator.
- Team Player and good Leadership skills.

Education

- *MBA at BizzNezz School.*
- *Bachelor of Science at Kalsh University.*

Professional Experience

- **2021 - Today:** *Cloud Admin at Storming Systems.*
- **2018 - 2021:** *IT Specialist at Loyal Bank.*
- **2017 - 2018:** *Trainee at Computer Land.*

Figure 4.2 – The resume returned by the cloud function

We can customize the header by passing the parameters, building a URL like this: `https://python-http-function-4slsbxpeoa-uc.a.run.app/?template=english.html&name=John&company=StarTalent`. And voilà, our cloud-based resume server is now online!

Of course, this sample cloud function could benefit from a lot of improvements. These are just some of the ideas I can think of, but I'm sure you will have many more:

- Encoded URLs, so that people won't see their name passed as a parameter, for example, using Base64 to hash the parameters and hide them in a URL like this: `https://hostname/?resumekey=dGVtcGxhdGU9ZW5nbGlzaC5odG1sJm5hbWU9Sm9obiZjb21wYW55PVN0YXJUYWxlbnQ=`. This sample URL contains the exact same parameters used in the preceding example and would display the resume customized for John Smith from StarTalent using the English template.

- Allow online real-time **What You See Is What You Get** (**WYSIWYG**) editing of the content of the resume, using GCS to store each different template and revision.

> **Note**
>
> Second-generation Cloud Functions URLs are non-deterministic at the time I'm writing this, but this is on the roadmap and is quite useful since you can *guess* the URL just by knowing the region, the name of the Google Cloud project, and the name of the cloud function.

Once a cloud function has been deployed, we can see a lot of information about it in the Google Cloud console. The **Cloud Functions** section can always be accessed using the direct link `https://console.cloud.google.com/functions/list`, or we can use the direct link that appears when we deploy a function, after the text **You can view your function in the Cloud Console here:**.

Once you open the preceding link, you will see a list of all the cloud functions deployed, including information such as the function name, the region, the runtime version, memory allocated, or which function is executed once it's triggered. At the end of each line on the list, there is an icon with three dots that allows quick access to see the logs, allows you to make a copy of a cloud function, and can also take you to the Cloud Run-associated service.

Clicking on a function name will take you to another screen with a tabbed interface. Let's summarize what you can see in each of those tabs since they can be extremely useful.

The **Metrics** tab will show you some interesting numbers about your cloud function, such as the number of invocations per second or the number of active instances, which is useful to see the traffic in real time.

In order to simulate real traffic, I used the `hey` command-line tool, which is available in Cloud Shell and can be used for this purpose. Just invoking it by passing a URL as a parameter will generate 200 requests, but you can customize it using many other options. For my test, I used this Bash one liner, which generates random traffic, and left it running. If you want to use it, just replace <YOUR-CLOUD-FUNCTION-URL> with the URL to your cloud function:

```
while sleep $[ ( $RANDOM % 300 )  + 1 ]s; do hey https://<YOUR-
CLOUD-FUNCTION-URL>/; done
```

These were the metrics I got after an hour:

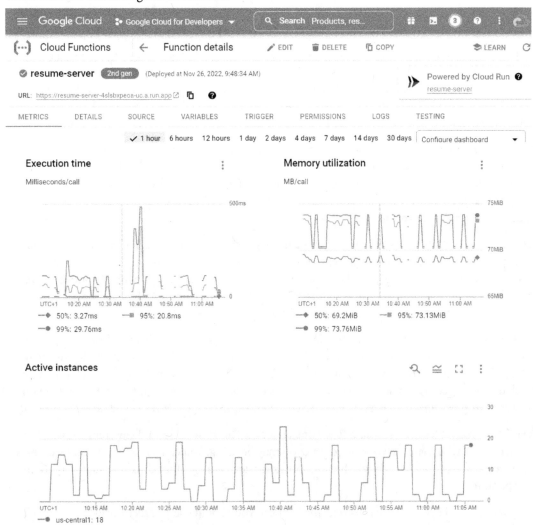

Figure 4.3 – The metrics information for our resume cloud function

The preceding graphs are interesting because you can see how Cloud Functions is scaling up when the 200 requests are received and then goes back to 0. You can also see how long requests take and how much memory is used: there is some variability in the numbers due to Cloud Run warming up after a cool-down. All this can help us better understand how our code runs, and what we can do to improve its behavior.

There are two quite useful metrics to look at: *execution time* and *memory utilization*. Execution time allows us to understand whether our cloud function is well designed and runs to completion before the configured timeout is triggered; otherwise, we should redeploy our function, increasing its value. In second-generation Cloud Functions, the maximum timeout duration is 60 minutes (3,600 seconds) for HTTP functions and 9 minutes (540 seconds) for event-driven functions.

Finally, memory utilization allows us to see how much memory our cloud functions are using in each execution. Each deployment configures the amount of memory and vCPU allocated to run, with a default of 256 MiB and .167 vCPU, a sixth part of a 2.4 GHz CPU. In the preceding screenshot, you can see that our resume-serving function uses less than 100 MiB in each run, so we could even decrease the amount of allocated memory to 128 MiB and save costs even more.

On the other end, if our cloud function tries to use more than the allocated amount of memory, it will crash, so this graph can help us detect this situation and redeploy the cloud function, this time increasing the amount of memory using the `--memory` flag. The list of available memory and CPU tiers is available on this documentation page: `https://cloud.google.com/functions/docs/configuring/memory`.

The second tab, **Details**, offers information about the deployment, including region, timeout, minimum, and maximum configured instances, service account, or networking settings. You can modify any of these parameters by redeploying the cloud function with an updated deployment command. You can see the full list of parameters on this documentation page: `https://cloud.google.com/sdk/gcloud/reference/functions/deploy`.

In the third tab, **Source**, you can see the deployed code, which is read-only by default. But you can also click on the **EDIT** button at the top and proceed to make changes on the fly and redeploy by just clicking the **Deploy** button once you are finished. Beware of using this option too frequently, since any changes applied locally may leave your deployed code out of sync with the one stored in your code repository. However, this option can still be a lifesaver if you need a very quick fix to be applied.

The next tab, **Variables**, contains all information about the **Runtime** and **Build** environment variables, together with any secrets that your cloud function may be using. You can pass environment variables to your cloud function, which will be read on execution time and used in a similar way to any other parameters.

If any of these variables include sensitive information, you should use secrets instead, which will be both safely stored and retrieved, thus reducing the chances of unwanted leaks.

You can pass environment variables using the `--set-build-env-vars=[KEY=VALUE,...]` build parameter or use `--build-env-vars-file=FILE_PATH` and pass the path to a YAML file including a key and value pair in each line. For secrets, you can use `--set-secrets=[SECRET_ENV_VAR=SECRET_VALUE_REF,/secret_path=SECRET_VALUE_REF,/mount_path:/secret_file_path=SECRET_VALUE_REF,...]` to define the secrets or `--update-secrets=[SECRET_ENV_VAR=SECRET_VALUE_REF,/secret_path=SECRET_VALUE_REF,/mount_path:/secret_file_path=SECRET_VALUE_REF,...]]` to update them.

The next tab, **Trigger**, allows us to see how we can trigger our cloud function, either using HTTP or an event, including a direct link to invoke our cloud function.

Then, the next one, **Permissions**, summarizes all security entries defined for our cloud function, both by **Principals** (who can do what) or by **Role** (which groups can do what). Unless a cloud function has been deployed enabling anonymous invocation, only users with invoke permission will be able to run a cloud function. You should be extremely careful and only open anonymous access if you really want your functions to be publicly accessible. Otherwise, your cloud function could be triggered by anyone, and this may not only lead to security issues but also increase your costs since you will be charged by the number of runs and online scanners and bots may increase your numbers and you may have a nasty surprise when the billing cycle ends.

> **Note**
>
> Properly securing your serverless deployments is a key step that you should include in your development cycles, sprints, or iterations.

There are just two more tabs in the interface: the first is the **Logs** tab, which provides us with access to the latest log entries. Personally, I prefer to open a link to the **Logs Viewer** on a new screen from the three-dots icon on the Cloud Functions list page, but this can also be useful to identify any recent issues when our cloud function fails.

Finally, the **Testing** tab can be useful for fast tests, since it will help us quickly build a payload and trigger our cloud function, so we can then switch back to the **Logs** tab and check that everything works as expected.

Tests can also be done using the command line, with a snippet like the one following this paragraph, which I used to test an HTTP function. This code is also available in the book's repository, so you can try it if you want to. Notice how, in the following screenshot, the `name` parameter is passed in the URL and the HTML response is customized for that specific name in the line containing *"(Specially prepared for...)"*. An authorization token is also included so that the testing is also compatible with Cloud Functions not allowing anonymous invocations:

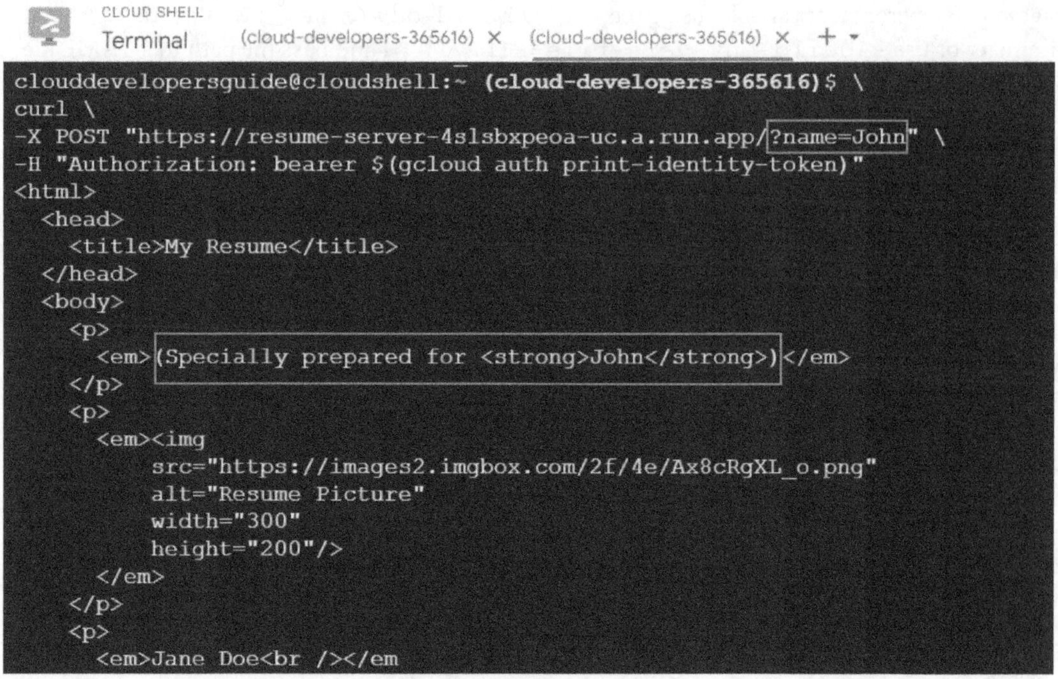

```
CLOUD SHELL
Terminal        (cloud-developers-365616)  ✕    (cloud-developers-365616)  ✕    +  ▾

clouddevelopersguide@cloudshell:~ (cloud-developers-365616)$ \
curl \
-X POST "https://resume-server-4slsbxpeoa-uc.a.run.app/?name=John" \
-H "Authorization: bearer $(gcloud auth print-identity-token)"
<html>
  <head>
    <title>My Resume</title>
  </head>
  <body>
    <p>
      <em>(Specially prepared for <strong>John</strong>)</em>
    </p>
    <p>
      <em><img
          src="https://images2.imgbox.com/2f/4e/Ax8cRgXL_o.png"
          alt="Resume Picture"
          width="300"
          height="200"/>
      </em>
    </p>
    <p>
      <em>Jane Doe<br /></em
```

Figure 4.4 – Using curl to test the cloud function from the command line

> **Note**
>
> The preceding example uses POST to send the payload data, and your code may only be ready to handle GET data. It's up to you whether you implement support for both methods or just one, depending on your use case and how the HTTP function will be invoked. GET exposes the parameters in the URL while POST sends them as data.

You can use similar commands to send Pub/Sub messages or to generate any other type of events in production, but it is a good practice to use emulators during our internal tests. You can read more about this topic in the following documentation section: https://cloud.google.com/functions/docs/testing/test-overview.

Debugging a cloud function

As I did in the section about testing, I will cover Cloud Functions debugging considering two different environments: local debugging before deployment and remote debugging, once our cloud function is running on Google Cloud.

In the first case, for local debugging, we can use the function's framework in debug mode to help us have a better understanding of what's happening during the execution of a cloud function. When the framework has started adding the --debug flag, it will log all requests and logging events to the local console, which can be very useful if we developed our application including a flag to enable verbose logging, so we can follow the execution by looking at the events and better understand where our code is failing.

Once we have deployed our cloud function in Google Cloud, we can also use Stackdriver to connect to the running container and be able to debug it. This process is not as simple as the others described, but it can be a lifesaver once you get used to it.

If you are interested in this topic, I added some detailed articles about testing and debugging Cloud Functions in the *Further reading* section, at the end of the chapter.

Tips and tricks for running your code using Cloud Functions

I have put together some tips for getting the most out of Cloud Functions.

Whenever it's possible, you should code cloud functions to minimize their memory usage. Since the cost is proportional to the memory and vCPU allocated, loading big structures in memory will increase costs. If you can implement your use case in a simpler way and not require processing huge amounts of data in each execution, Cloud Functions will be not only the best technical fit but also the most affordable one.

Also, if you need to test your cloud function in production, you can directly execute it from the console using the gcloud utility as follows:

```
gcloud functions call python-http-function --data
'{"template":"english.html", "name":"John Smith",
"company":"StarTalent"}'
```

As you will have noticed, this is an example of testing our resume server. Please take into account that when using the gcloud command, the data is passed using POST, so you may need to make some small changes in the source code to also support this method besides GET.

You can read more about direct invocation on the following documentation page: https://cloud.google.com/functions/docs/running/direct.

Another interesting tip is that we can deploy a cloud function directly from source code located in Cloud Source Repositories, also enabling us to use GitHub or Bitbucket, thus reducing the complexity of our development and testing workflows because we no longer need to check out our code before deploying it.

For this to work, we must first set up a repository and, optionally, connect it to our GitHub or Bitbucket repository if we want to use them as sources. Then, we can use the following URL in the --source flag:

```
https://source.developers.google.com/projects/PROJECT_ID/repos/
REPOSITORY_NAME
```

You can read more about this feature and other advanced deployment techniques on this page of the official documentation: `https://cloud.google.com/functions/docs/deploy`

How much does it cost to run a cloud function?

One of the most common questions about running our code on any cloud provider is how to accurately estimate how much we will be paying at the end of the month. While this can be easy for some services, it becomes more complicated with others, as is the case with Cloud Functions.

The fact that a cloud function is just a piece of architecture, meaning that it will be using other services most of the time, complicates finding an answer.

For example, imagine that you have a thumbnail generation service where you schedule your cloud function to run every 10 minutes using Cloud Scheduler and use it to get information about the latest image uploads by querying a BigQuery table, and then loading the image from GCS, finally generating and storing a thumbnail in another directory of the bucket.

This small example has made use of the following Google Cloud services:

- Cloud Functions
- Cloud Scheduler
- BigQuery
- GCS

So, you will be charged for each of them, depending on your usage and whether you exceeded the free tier, where available. Please take this into account and make sure you consider any external services that you are using when you try to calculate your total costs.

Considering just Cloud Functions, you will highly likely incur the following charges:

- **Artifact Registry**: For storing your function (but this has a free tier).
- **Cloud Build**: Used to build an executable image containing your function.
- **Eventarc**: Used for event delivery.
- **Cloud Functions invocations**: Charged by every million invocations, with the first two included in the free tier.
- **Compute time**: There is a free tier, too.
- **Internet egress traffic**: Charged by GB. The first 5 GB are included in the free tier.

As you can see, cost calculations can become complicated, and that's why the detailed documentation page for Cloud Functions pricing (`https://cloud.google.com/functions/pricing`) includes a couple of real-world examples that can be useful to estimate costs in different scenarios:

- A simple event-driven function with 128 MB of memory and a 200 MHz CPU, invoked 10 million times per month and running for 300 ms each time using only Google APIs (no billable egress), will cost $7,20 every month

- A medium-complexity HTTP function with 256 MB of memory and a 400 MHz CPU, invoked 50 million times per month via HTTP, running for 500 ms each time and sending 5 KB of data back to the caller (5 KB egress per invocation) will cost $159.84

In my own experience, using Cloud Functions has always been an affordable option for short and repetitive operations, and, even with millions of invocations every month, I never saw costs over a few hundred dollars a month. Also, since the cost is proportional to the number of invocations, Cloud Functions can be an interesting alternative for services where more invocations also mean more revenue, so that monthly costs are just a small fraction of the benefits they provide.

However, there are other alternatives for running your code where costs can be more stable and predictable. Let's introduce App Engine.

Using App Engine to run your code

Now, it's time to move from FaaS to PaaS and introduce the second option for our serverless deployments: App Engine.

Introducing App Engine

App Engine (`https://cloud.google.com/appengine`) is a serverless PaaS product for developing and hosting our web applications. We can choose among many popular programming languages and use any framework or library to build our application, and Google Cloud will handle the infrastructure, including a demand-based scaling system to ensure that you always have enough capacity for our users.

This product is a very good fit for microservices-based architectures and requires zero server management and zero configuration deployment tasks, so we can focus on developing amazing applications. Indeed, we can use App Engine to host different versions of our app and use this feature to create separate environments for development, testing, staging, and production.

It's important that you know that there can only be one App Engine instance in each Google Cloud project and that whatever region you choose when you create it will become permanent, so please make that choice wisely.

An App Engine application (`https://cloud.google.com/appengine/docs/legacy/standard/python/an-overview-of-app-engine`) is made up of one or more services, each of which can use different runtimes, each of which can have customized performance settings. Each of our services can have multiple versions deployed that will run within one or more instances, depending on the amount of traffic that we configured it to handle.

All the resources of an application will be created in the region that we choose when we create our App Engine app, including code, settings, credentials, and all the associated metadata. Our application will include one or more services but must have at least what is called the default service, which can also have multiple versions deployed.

Each version of a service that we deploy in our app will contain both the source code that we want to run and the required configuration files. An updated version may contain changes in the code, the configuration, or both, and a new version will be created when redeploying the service after making changes to any of these elements.

The ability to have multiple versions of our application within each service will make it easier to switch between versions for cases such as rollbacks or testing and can also be very useful when we are migrating our service, allowing us to set up traffic splits to test new versions with a portion of the users before rolling them out to all of them.

The different deployed versions of our services will run on one or more instances depending on the load at each time. AppEngine will scale our resources automatically, up if required to maintain the performance level, or down to avoid resource waste and help reduce costs.

Each deployed version of a service must have a unique name, which can be used to target and route traffic to a specific resource. These names are built using URLs that follow this naming convention:

```
https://<VERSION>-dot-<SERVICE>-dot-<PROJECT_ID>.<REGION_
ID>.r.appspot.com
```

> **Note**
>
> The maximum length of `<VERSION>-dot-<SERVICE>-dot-<PROJECT_ID>` is 63 characters, where `VERSION` is the name of our version, `SERVICE` is the name of our service, and `PROJECT_ID` is our project ID, or a DNS lookup error will occur. Another limitation is that the name of the version and the service cannot start or end with a hyphen. Any requests that our application receives will be routed only to those versions of our services that have been configured to handle the traffic. We can also use the configuration to define which specific services and versions will handle a request depending on parameters such as the URL path.

App Engine environment types

App Engine offers two different environment types.

The **App Engine standard environment** is the simplest offering, aimed at applications running specific versions of the supported programming languages. At the time of writing this chapter, you can write your application in Node.js, Java, Ruby, C#, Go, Python, or PHP. You can see the up-to-date list on this documentation page: `https://cloud.google.com/appengine/docs/the-appengine-environments`.

In a standard environment, our application will run on a lightweight instance inside of a sandbox, which means that there will be a few restrictions that you should consider. For example, we can only run a limited set of binary libraries, restricting access to external Google Cloud services only to those available using the App Engine API, instead of the standard ones. Other particularly important limitations are that App Engine standard applications cannot write to disk and that the options of CPU and memory to choose from are limited. For all these reasons, App Engine standard is a genuinely precise fit for stateless web applications that respond to HTTP requests quickly, that is, microservices.

App Engine standard is especially useful in scenarios with sudden changes in traffic because this environment can scale very quickly and supports scaling up and down. This means that it can scale up your application quickly and effortlessly to thousands of instances to handle sudden peaks, and scale it down to zero if there is no traffic for some time.

If the mentioned limitations are not a problem for your use case, this can be a remarkably interesting choice to run your code, because you will pay close to nothing (or literally nothing).

App Engine standard instances are charged based on **instance hours**, but the good news is that all customers get 28 instances in a standard environment free per day, not charged against our credits, which is great for testing and even for running a small architecture virtually for free.

The second type is called the **App Engine flexible** environment. This one will give us more power, more options... and more responsibilities, at a higher cost. In this case, our application will be containerized with Docker and run inside a virtual machine. This is a perfect fit for applications that are expecting a reasonably steady demand and need to scale more gradually. The cons of this environment are that the minimum number of instances in App Engine flexible is 1 and that scaling in response to traffic will take significantly longer in comparison with standard environments.

On the list of pros, flexible environments allow us to choose any Compute Engine machine type to run our containerized application, which means that we have access to many more combinations of CPU, memory, and storage than in the case of a standard environment.

Besides, flexible environments have fewer requirements about which versions of the supported programming languages we can use, and they even offer the possibility of building custom runtimes, which we can use to add support for any other programming languages or versions that we may specifically need. This will require additional effort to set it up but also opens the door to running web applications written in any version of any language.

Flexible App Engine instances are billed based on **resource usage**, including vCPU, memory, and persistent disks.

Finally, most of the restrictions that affect App Engine standard instances do not apply to flexible environments: we can write to disk, use any library of our choice, run multiple processes, and use standard cloud APIs to access external services.

> **Note**
>
> Standard and flexible App Engine environments should not be mutually exclusive, but complementary. The idea is that we run simple microservices using fast scaling and cost-efficient standard environments whenever possible and complement them with flexible environments used for those microservices that will not work under the limitations of a standard environment. Specific requirements such as needing more CPU or memory, requiring disk access, or making API calls to use cloud services will justify the use of flexible instances. When combining both instance types, inter-service communication can be implemented using Pub/Sub, HTTP, or Cloud Tasks, which makes App Engine a great choice to create architectures combining always-on and on-demand microservices.

You can read an interesting comparison table detailing the similarities and key differences between both environment instances in the following documentation section: `https://cloud.google.com/appengine/docs/flexible/flexible-for-standard-users`.

Scaling strategies in App Engine

App Engine applications are built on top of one or more instances, which are isolated from one another using a security layer. Received requests are balanced across any available instances.

We can choose whether we prefer a specific number of instances to run despite the traffic, or we can let App Engine handle the load by creating or shutting down instances as required. The **scaling strategy** can be customized using a configuration file called `app.yaml`. Automatic scaling will be enabled by default, letting App Engine optimize the number of idle instances.

The following is a list of the three different scaling strategies available for App Engine:

- **Manual scaling**: A fixed number of instances will run despite changes in the amount of traffic received. This option makes sense for complex applications using a lot of memory and requiring a fast response.

- **Basic scaling**: As its name suggests, this option will make things simple by creating new instances when requests are received and shutting them down when instances have been idle for some time. This is a nice choice for applications with occasional traffic.

- **Automatic scaling**: This is the most advanced option, suitable for applications needing to fine-tune their scaling to prevent performance issues. Automatic scaling will let us define multiple metrics with their associated thresholds in our YAML configuration file. App Engine will use these metrics to decide when it's the best time to create new instances or shut down idle ones so that there is no visible effect on performance. We can also optionally use the `automatic_scaling` parameter to define the minimum number of instances to always keep running.

You can find a table comparing these scaling strategies in the documentation page about App Engine instance management: `https://cloud.google.com/appengine/docs/legacy/standard/python/how-instances-are-managed`

The differences between the different strategies are quite simple to explain. In basic scaling, App Engine prioritizes cost savings over performance, even at the expense of increasing latency and hurting performance in some scenarios, for example, after it scales to 0. If low latency is an important requirement for your application, this option will probably not work for you.

Automatic scaling, however, uses an individual queue for each instance, whose length is periodically monitored and used to detect traffic peaks, deciding when new instances should be created. Also, instances with queues detected to be empty for a while will be turned off, but not destroyed, so they can be quickly reloaded if they are needed again later. While this process will reduce the time needed to scale up, it may still increase latency up to an unacceptable level for some users. However, we can mitigate this side effect by specifying a minimum number of idle instances to always keep running, so we can handle sudden peaks without seeing our performance hurt.

Using App Engine in microservice architectures

When we build an application using a microservice architecture, each of these microservices implements **full isolation of code,** which means that the only communication method that we can use to execute their code is using HTTP or a RESTful API call. One service will otherwise never be able to directly execute code running on another. Indeed, it's common that different services are written using different programming languages too. Besides, each service has its own custom configuration, so we may be combining multiple scaling strategies.

However, there are some App Engine resources, such as **Cloud Datastore**, **Memcached**, or **Task Queues**, which are shared between all services running in the same App Engine project. While this may have advantages, it may be a risk since a microservices-based application must maintain code and data isolation between its microservices.

While there are some architectural patterns that can help mitigate unwanted sharing, enforced separation can be achieved by using multiple App Engine projects at the expense of worse performance and more administrative overhead. A hybrid approach can also be a very valid option.

The App Engine documentation contains more information about microservices, including a comparison of service and project isolation approaches, so you can make a better choice for your architecture: `https://cloud.google.com/appengine/docs/legacy/standard/python/microservices-on-app-engine`.

Before getting to the example, let's introduce configuration files, which are key for deploying App Engine applications.

Configuring App Engine services

Each version of an App Engine service has an associated `.yaml` file, which includes the name of the service and its version. For consistency, this file usually takes the same name as the service it defines, while this is not required. When we have multiple versions of a service, we can create multiple YAML files in the same directory, one for each version.

Usually, there is a separate directory for each service, where both its YAML and the code files are stored. There are some optional application-level configuration files, such as `dispatch.yaml`, `cron.yaml`, `index.yaml`, and `queue.yaml`, which will be located in the top-level directory of the app. However, if there is only one service or multiple versions of the same service, we may just prefer to use a single directory to store all configuration files.

Each service's configuration file is used to define the configuration of the scaling type and instance class for a specific combination of service and version. Different scaling parameters will be used depending on the chosen scaling strategy, or otherwise automatic scaling will be used by default.

As we mentioned earlier, the YAML can also be used to map URL paths to specific scripts or to identify static files and apply a specific configuration to improve the overall efficiency.

There are four additional configuration files that control optional features that apply to all the services in an app:

- `dispatch.yaml` overrides default routing rules by sending incoming requests to a specific service based on the path or hostname in the URL
- `cron.yaml` configures regularly scheduled tasks that operate at defined times or regular intervals
- `index.yaml` specifies which indexes your app needs if using Datastore queries
- `queue.yaml` configures push and pull queues

After covering all the main topics related to App Engine, it's time to deploy and run some code to see all the discussed concepts in action.

Writing, deploying, and running code with App Engine

We will now deploy our resume-serving application in App Engine and see the differences between this implementation and the one using Cloud Functions.

The first file that we will create for our application is `app.yaml`, which can be used to configure a lot of settings. In our case, it will include the following contents:

```
runtime: python38
service: resume-server

handlers:
- url: /favicon\.ico
  static_files: favicon.ico
  upload: favicon\.ico
```

First, we will define which runtime we want to use. In this case, it will be a Python 3.8 module. Then, we will define a service name. I chose `resume-server` just in case you were already using the `default` service for any other purposes. Please remember that if this parameter is not defined in the file, the app will be deployed to the `default` service.

Since App Engine is a full application server, I'm taking the chance to include a favicon, that is, an icon that the web browser will show next to the page title. In this case, we just add the icon file, called `favicon.ico`, and add a rule to serve the icon when it is requested. The runtime will forward the rest of the requests by default to a file called `main.py`, which will be the next file that we will talk about.

As its name may suggest, `main.py` contains the core of the code and it is indeed quite similar to the version that we created as a cloud function. There are some differences at the beginning of the file because we will be using Flask to handle the requests and an instance of Cloud Logging when the app is deployed in production:

```
from flask import request, current_app, Flask
from google.cloud import storage
import google.cloud.logging
import logging

BUCKET_NAME = "<YOUR_BUCKET_NAME>"
DEFAULT_TEMPLATE_NAME = "english.html"

app = Flask(__name__)
app.debug = False
app.testing = False

# Configure logging
if not app.testing:
```

```
logging.basicConfig(level=logging.INFO)
client = google.cloud.logging.Client()
# Attaches a Cloud Logging handler to the root logger
client.setup_logging()
```

After these lines, you will see the same functions that we already covered earlier in this chapter, until we get to the last few lines of the file. Notice how now we have one line for routing requests to the root URL and how the last line runs the app, making it listen on the loopback interface for local executions:

```
DEFAULT_TEMPLATE = "english.html"
@app.route('/')
def get():
    template = request.args.get('template', DEFAULT_TEMPLATE)
    name = request.args.get('name', None)
    company = request.args.get('company', None)
    resume_html = return_resume(template, name, company)
    return resume_html


# This is only used when running locally. When running live,
# gunicorn runs the application.
if __name__ == '__main__':
    app.run(host='127.0.0.1', port=8080, debug=True)
```

The deployment package also includes a requirements.txt file. In this case, these are its contents:

```
Flask==2.2.2
google-cloud-storage==2.5.0
google-cloud-logging==3.2.4
```

Notice how all three imported packages have their version frozen, for the sake of stability in future deployments, as we already discussed.

Now we are ready for testing, and the four files have been copied to the same working directory: app.yaml, favicon.ico, main.py, and requirements.txt.

Python's virtualenv and pytest can be used for local fast testing, and they are indeed recommended as the first option, rather than using dev_appserver, which is the local development server that Google Cloud SDK provides. However, if you are still interested, there's information about it in this section of the official documentation: https://cloud.google.com/appengine/docs/standard/tools/using-local-server.

Please notice that simulated environments may not have exactly the same restrictions and limitations as the sandbox. For example, available system functions and runtime language modules may be restricted, but timeouts or quotas may not.

The local development server will also simulate calls to services such as Datastore, Memcached, and task queues by performing their tasks locally. When our application is running in the development server, we can still make real remote API calls to the production infrastructure using the Google API's HTTP endpoints.

Another option to simulate a production App Engine environment is to use a **Web Server Gateway Interface (WSGI)** server locally by installing gunicorn in Cloud Shell using the following command:

```
pip install gunicorn
```

Then, we will just run it using our app as an entry point, as in the following example:

```
gunicorn -b :$PORT main:app
```

Here, $PORT is the port number we defined for our application, 8080 by default, and main:get is the name of the Python file and the function to execute when a request is received.

In my example, I invoked it using the following command line in Cloud Shell, so that it runs in the background:

```
/home/<user>/.local/bin/gunicorn -b :8080 main:app &
```

Now, we can send requests using curl and validate the output as part of our unit tests. For example, our usual test URL would now be triggered using the following command. Please don't forget the double quotes, or otherwise only the first parameter will be received:

```
curl "http://127.0.0.1:8080/?template=english.
html&name=John&company=StarTalent"
```

Applications designed for flexible environments can also be directly executed for testing, given that they will have direct access to cloud services. Using emulators is often recommended in cases like this in order to avoid incurring excessive costs while running the tests.

After successfully passing all local tests, the application will be ready for deployment. And it couldn't be any simpler than running the following command in the console from the working directory containing all the files previously mentioned:

```
gcloud app deploy app.yaml
```

This deployment command supports other flags, such as the following:

- `--version` to specify a custom version ID
- `--no-promote` to prevent traffic from being automatically routed to the new version
- `--project` to deploy to a specific Google Cloud project

As it happened with Cloud Functions, you may be asked to authenticate yourself during the deployment, and you could also be asked to enable any required APIs the first time that you deploy an app. In the case of App Engine, this is an example of the output of a deployment command:

```
Services to deploy:
descriptor:                    [/home/clouddevelopersguide/App_
Engine/app.yaml]
source:                        [/home/clouddevelopersguide/App_
Engine]
target project:                [cloud-developers-365616]
target service:                [resume-server]
target version:                [20221021t201413]
target url:                    [http://resume-server.cloud-
developers-365616.uc.r.appspot.com]
target service account:        [App Engine default service
account]


Do you want to continue (Y/n)? Y


Beginning deployment of service [resume-server]...
Uploading 1 file to Google Cloud Storage
100%
100%
File upload done.
Updating service [resume-server]...done.
Setting traffic split for service [resume-server]...done.
Deployed service [resume-server] to [http://resume-server.
cloud-developers-365616.uc.r.appspot.com]


You can stream logs from the command line by running:
  $ gcloud app logs tail -s resume-server
```

```
To view your application in the web browser run:
  $ gcloud app browse -s resume-server
```

Notice how we can use the last section of the output to get the URL to the application, and the one right above it to print the logs in the command-line console. The deployment process involves copying our files to GCS, and then updating the service and setting its traffic split.

Once we obtain the URL, we can again append the parameters to test the application. In my case, this was the complete URL:

```
https://resume-server-dot-cloud-developers-365616.uc.r.appspot.
com/?template=english.html&name=John+Smith&company=StarTalent
```

You can read more about testing and deploying your applications in App Engine in this section of the official documentation site: `https://cloud.google.com/appengine/docs/standard/testing-and-deploying-your-app`.

Debugging in App Engine

Luckily for us, App Engine is compatible with many of the tools that we already introduced for testing and debugging our cloud functions. With App Engine, we can also use Cloud Monitoring and Cloud Logging to monitor the health and performance of our app, and **Error Reporting** to diagnose and fix bugs quickly. Cloud Trace can also help us understand how requests propagate through our application.

Cloud Debugger can help us inspect the state of any of our running services without interfering with their normal behavior. Besides, some IDEs, such as IntelliJ, allow debugging App Engine standard applications by connecting to a local instance of `dev_appserver`. You can find more information in this section of the official documentation site: `https://cloud.google.com/code/docs/intellij/deploy-local`.

After completing the whole development cycle when using App Engine, it's the perfect time to explain how we will be billed if we decide to use App Engine.

How much does it cost to run your code on App Engine?

App Engine pricing scales with our app's usage, and there are a few basic components that will be included in the App Engine billing model, such as standard environment instances, flexible environment instances, and App Engine APIs and services.

As I mentioned earlier, flexible App Engine instances are billed based on resource utilization, including vCPU, memory, persistent disks, and outgoing network traffic. Standard App Engine instances follow a much simpler model based on the number of hours they have been running for. Any other APIs and services used should be also added to the bill, such as Memcached, task queue, or the Logs API.

For more details about the pricing, you can refer to this documentation section: `https://cloud.google.com/appengine/pricing`

Regarding the free tier, users get 28 standard frontend instances and 9 backend instances for free every day, and new customers get $300 in free credits to spend on App Engine. You may find all the details about quotas in the following section of the documentation website: `https://cloud.google.com/appengine/docs/standard/quotas`.

To get an estimate of our bill, we can use the Google Cloud Pricing Calculator available in the following section of the documentation: `https://cloud.google.com/products/calculator#tab=app-engine`.

Tips and tricks for running your code on App Engine

If you read the *Limits* section at the end of the *App Engine Overview* section of the documentation (`https://cloud.google.com/appengine/docs/legacy/standard/python/an-overview-of-app-engine`), you will see that there are different limits for the number of services and instances depending on the application type (free or paid) and whether the app is hosted in *us-central* or in any other region. You should take these numbers into account when you decide which application type to use.

If our app uses automatic scaling, it will take approximately 15 minutes of inactivity for the idle instances to start shutting down. To keep one or more idle instances running, we should set the value of `min_idle_instances` to 1 or higher.

Regarding security, a component called the App Engine firewall can be used to set up access rules. Managed SSL/TLS certificates are included by default on custom domains at no additional cost.

This was all the information we need to know about App Engine. Now, it's time to wrap up.

Summary

In this chapter, we discussed how Cloud Functions and App Engine work, what their requirements are, and how much they cost. We also covered how we can use them to run our code and how we can test and troubleshoot our applications and services when they use these products. Finally, we implemented the same example using both options.

In the next chapter, we will cover Cloud Run, the third option to run serverless code on Google Cloud using a CaaS model.

Further reading

To learn more about the topics that were covered in this chapter, take a look at the following resources:

- *What is virtualization?* (https://www.redhat.com/en/topics/virtualization/what-is-virtualization)

- *What are Cloud Computing Services [IaaS, CaaS, PaaS, FaaS, SaaS]* (https://medium.com/@nnilesh7756/what-are-cloud-computing-services-iaas-caas-paas-faas-saas-ac0f6022d36e)

- *How to Develop, Debug and Test your Python Google Cloud Functions on Your Local Development Environment* (https://medium.com/ci-t/how-to-develop-debug-and-test-your-python-google-cloud-functions-on-your-local-dev-environment-d56ef94cb409)

5

Running Serverless Code on Google Cloud – Part 2

After covering Cloud Functions and App Engine in the previous chapter, this one will introduce the third serverless option to run our code on Google Cloud, this time using containers: Cloud Run.

First, I will introduce its basic concepts and describe the two different execution environments available. Then, we will see together how we can run our code using Cloud Run and what the best practices for debugging it are. Next, I will show you how much Cloud Run costs and include some tips and tricks to help you get the most out of it.

Finally, we will discuss the similarities and differences between the three serverless products covered in this and the previous chapter, so you can better decide when you should use each.

We'll cover the following main topics in this chapter:

- Using Cloud Run to run your code
- Choosing the best serverless option for each use case

Let's get started!

Using Cloud Run to run your code

Cloud Run is the third and last option for serverless code execution that we will discuss in this chapter. This is the CaaS option, and you should keep an eye on it because, as we will see in the next couple of chapters, containers are the biggest bet for portable development as of today, and the base for Google's hybrid and multi-cloud offering.

Introducing Cloud Run

Cloud Run (`https://cloud.google.com/run/docs/overview/what-is-cloud-run`) is the third serverless option for running our code on Google Cloud. In this case, our code will be running on containers on top of Google's scalable infrastructure, once again forgetting about everything to do with operational tasks or scaling our architecture, since Google's CaaS compute platform will take care of it for us.

This also means that Google will decide when to stop sending requests and even when to terminate an instance, and since each of them will be ephemeral and disposable, our code should be well prepared for *imminent disposal*.

Cloud Run offers some interesting features for developers, which can help us accommodate specific use cases very easily:

- First, we can run virtually any kind of code written in our preferred programming language, using Cloud Run, as long as it can be containerized. For our code to work, we can either build our own container image, using any programming language, as long as we include all libraries and dependencies and even binaries if we need them, or we can use a feature called **source-based deployment**, where Cloud Run will build the container image for us, ready to run code written on one of the supported languages: Go, Node.js, Python, Java, Kotlin, .NET, or Ruby.

- Second, since we will be charged based either on the number of requests it served or the resources it used while running, we can run it continuously as a service, so it can respond to HTTP requests or events but may be idling quite often, or we can run it as a job, meaning that it will perform a specific task and then quit once it has finished.

The service option can help us save costs and it can be a great fit both for single-use tasks, such as a migration or installation job, and for repetitive maintenance tasks, such as data clean-ups, daily aggregations, or similar scenarios that are scheduled to run periodically, but not too frequently.

In order to be a good fit for Cloud Run, our application will need to meet all the following criteria:

- Either it serves requests, streams, or events delivered via HTTP, HTTP/2, WebSockets, or gRPC, or it executes to completion
- Does not require a local persistent filesystem
- It's built to handle multiple instances of the app running simultaneously
- Does not require more than 8 CPUs and 32 GiB of memory per instance
- Meets one of the following criteria:
  - Is containerized
  - Is written in Go, Java, Node.js, Python, or .NET

We can otherwise containerize it.

Now, let's take a look at some of the basic concepts that can help us understand how Cloud Run works.

Basic concepts of Cloud Run

Looking at Cloud Run's resource model (`https://cloud.google.com/run/docs/resource-model`), there are some interesting concepts that we should be familiarized with before we start working on some examples:

A **service** is the main resource of Cloud Run. Each service is located in a specific **Google Cloud Platform** (**GCP**) region. For redundancy and failover, services are automatically replicated across multiple zones in that same region. Each service exposes a unique HTTPS endpoint on a unique subdomain of `*.run.app domain` and automatically scales the underlying infrastructure to handle incoming requests.

Similarly to what we just mentioned in the case of App Engine, each new deployment of a service in Cloud Run makes a new **revision** of that service to be created, which includes a specific container image, together with configuration settings such as environment variables or memory limits. Revisions are immutable, so any minor change will create a new revision, even if the container image remains intact and only an environment variable was updated.

We must take into account the following requirements when we develop a service using Cloud Run:

- The listening port must be customizable using the **PORT environment variable**. Our code will be responsible for detecting the optional use of this variable and updating the port used to listen for requests, in order to maximize portability. The service must be stateless. It cannot rely on a persistent local state.

- If the service performs background activities outside the scope of request handling, it must use the **CPU always allocated** setting.

- If our service uses a network filesystem, it must use the second-generation execution environment.

Regarding concurrency (`https://cloud.google.com/run/docs/about-concurrency`), Cloud Run behaves similarly to App Engine with autoscaling: each revision will be automatically scaled to the number of container instances needed to handle the queue of pending requests, but will also be scaled down if there is less or no traffic. Indeed, Cloud Run is also a **zero-scaling service**, which means that, by default, it will dispose of even the last remaining instance if there is no traffic to serve for a specific amount of time. We can change this behavior and eliminate cold starts by using the **min-instance** setting at the expense of increasing costs. You can read more details about how autoscaling works for Cloud Run at this link: `https://cloud.google.com/run/docs/about-instance-autoscaling`.

> **Note**
>
> A container instance can receive many requests at the same time, and this will lead to more resources being used, which also will mean higher costs. To give us more control over the limits of the scaling process, we can set the maximum number of requests that can be sent in parallel to a given container instance.

By default, each Cloud Run container instance can receive up to 80 requests at the same time. This is the maximum number of requests, and other metrics will be considered, such as CPU usage, to decide the final number, which could be lower than this maximum.

If our microservice can handle more queries, we can increase this to a maximum of 1,000. Although it is recommended to use the default value, we can also lower it in certain situations. For example, if our code cannot process parallel requests or if a single request will need to use all the CPU resources allocated, we can set the concurrency to 1 and our microservice will then only attend to one request at a time.

Requests are routed by default to the latest **healthy service revision** as soon as possible. The health of a service is probed by the load balancer, and a revision may be marked as unhealthy if it does not respond successfully a separately configurable number of times. For this reason, testing and debugging each service properly is especially important, in order to properly configure startup and liveness probes to detect when a service is not starting or suddenly stops working, so that it can be automatically restarted, thus reducing downtime.

We can also split traffic to multiple revisions at the same time, in order to reduce the risk while deploying a new revision. We can start by sending 1% of requests to a new revision and increase that percentage progressively while closely testing that everything works as expected until we complete the rollout, with a final scenario where 100% of the requests are sent to the latest revision.

As I mentioned previously, Cloud Run also supports the execution of jobs. Each **Cloud Run job** runs in a specific Google Cloud region and consists of one or multiple independent tasks that are executed in parallel in each job execution. Each task runs one container instance to completion and might retry it if it fails. All tasks in a job execution must complete for the job execution to be successful.

We can set timeouts for the tasks and even specify the number of retries in case of failure. If any task exceeds its maximum number of retries, it will be marked as failed and the parent job will be, too. By default, tasks execute in parallel up to a maximum of 100, but we can define a lower maximum if the level of usage of resources requires it.

Besides, Cloud Run introduces the concept of **array jobs**, where repetitive tasks within a job can be split among different instances to be run in parallel, thus reducing the time required for the full job to complete. This turns App Engine into a remarkably interesting choice if we need to process objects in batches, such as cropping lots of images, translating an extensive list of documents, or processing a big set of log files.

Considering the concepts that we just introduced, there are different options for triggering the execution of our code on Cloud Run. Let's enumerate them and provide a link with more information about each:

- Using HTTPS requests (`https://cloud.google.com/run/docs/triggering/https-request`)

- Using gRPC to enjoy the benefits of protocol buffers (`https://cloud.google.com/run/docs/triggering/grpc`)

- Using WebSockets (`https://cloud.google.com/run/docs/triggering/websockets`)

- Using Pub/Sub push (`https://cloud.google.com/run/docs/triggering/pubsub-push`)

- Using Cloud Scheduler to run services at a specific time (`https://cloud.google.com/run/docs/triggering/using-scheduler`)

- Using Cloud Tasks to execute them asynchronously (`https://cloud.google.com/run/docs/triggering/using-tasks`)

- Using Eventarc events as triggers (`https://cloud.google.com/run/docs/triggering/trigger-with-events`)

- Using workflows as a part of a pipeline (`https://cloud.google.com/workflows`)

Now that we have covered the basic concepts, let's take a look at the two execution environments that Cloud Run provides.

The two different execution environments to choose from

As is the case with Cloud Functions, Cloud Run has two different generations of execution environments (`https://cloud.google.com/run/docs/about-execution-environments`). Cloud Run services, by default, operate within the first-generation execution environment, which features fast cold-start times and emulation of most, but not all, operating system calls.

Originally, this was the only execution environment available to services in Cloud Run. This generation is the best choice for either bursty traffic that requires scaling out fast, or for the opposite case with infrequent traffic where our service frequently scales out from 0. If our services use less than 512 MiB of memory, which is the minimum for second-generation instances, we may also benefit from cost savings by choosing the first generation.

The second-generation execution environment for Cloud Run instances provides faster CPU and network performance, the latter especially in the presence of packet loss, and full Linux compatibility instead of system call emulation, including support for all system calls, namespaces, and cgroups, together with the support of network filesystem. These features make it the best choice for steady traffic, where scaling and cold starts are much less frequent, and for services that are intensive in CPU usage or make use of any of the new specific features provided by this generation.

Second-generation Cloud Run instances are in the preview phase at the time of writing this, and while the second-generation execution environment generally performs faster under sustained load, it has longer cold-start times than the first generation, which is something to consider when making a choice, depending on the specifics of your application or service.

We can specify the execution environment for our Cloud Run service when we deploy either a new service or a new revision of it. If we don't specify an execution environment, the first generation is used by default. Cloud Run jobs, however, automatically use second-generation execution environments, and this cannot be changed in the case jobs.

Now that we are done with the basic concepts of Cloud Run, let's move on and clarify a few requirements before we start developing a Cloud Run example service.

Writing and running code using Cloud Run

There are a few requirements for our Cloud Run services that are included in the **container runtime contract** (`https://cloud.google.com/run/docs/container-contract`) and we should take them into consideration before actually starting to write our code.

The code running in our Cloud Run container must listen for requests on IP address `0.0.0.0` on the port to which requests are sent. By default, requests are sent to port `8080`, but we can configure Cloud Run to send requests to another port of our choice, as long as it is not already in use.

Cloud Run injects the **PORT environment variable** into the container. Inside Cloud Run container instances, the value of the PORT environment variable always reflects the port to which requests are sent. Again, it defaults to `8080`.

A particularly important note to make, and a common reason for early errors among beginners, is, *our container should not implement any* **Transport Layer Security (TLS)** *directly*. TLS is terminated by Cloud Run for HTTPS and **gRPC**. Then, requests are proxied as HTTP/1 or gRPC to the container without TLS. If you configure a Cloud Run service to use HTTP/2 from end to end, your container must handle requests in HTTP/2 cleartext (`h2c`) format because TLS is still ended automatically.

For Cloud Run services, our container instance must send a response within the time specified in the **request timeout setting** after it receives a request, including the container instance startup time. Otherwise, the request is ended and a `504` error is returned.

With these requirements in mind, it's time to start building our first Cloud Run service, which we will use to implement our resume server using a container.

First of all, we will create a file called `.dockerignore`, which will contain a list of patterns of filenames that will not be copied to the container in any case:

```
Dockerfile
README.md
*.pyc
*.pyo
*.pyd
__pycache__
.pytest_cache
```

Then, we will use the slim Dockerfile template, which contains the following lines:

```
# Use the official lightweight Python image.
# https://hub.docker.com/_/python
FROM python:3.10-slim


# Allow statements and log messages to immediately appear in
the Knative logs
ENV PYTHONUNBUFFERED True


# Copy local code to the container image.
ENV APP_HOME /app
WORKDIR $APP_HOME
COPY . ./


# Install production dependencies.
RUN pip install --no-cache-dir -r requirements.txt


# Run the web service on container startup. Here we use the
# gunicorn webserver, with one worker process and 8 threads.
# For environments with multiple CPU cores, increase the
# number of workers to be equal to the cores available.
# Timeout is set to 0 to disable the timeouts of the workers
# to allow Cloud Run to handle instance scaling.
CMD exec gunicorn --bind :$PORT --workers 1 --threads 8
--timeout 0 main:app
```

Since we will be running our application from a container, we must include the resume template directly in the working directory, in this case, called english.html.

There will also be a requirements.txt file, but this time the number of lines will be shorter:

```
Flask==2.2.2
gunicorn==20.1.0
```

Finally, main.py will be very similar to other versions, so let's just take a look at the last few lines of code. As you will see, the main difference is that parameters are now passed as environment variables:

```
DEFAULT_TEMPLATE_NAME = "english.html"
@app.route('/')
```

```python
def get():
    template = request.args.get('template', DEFAULT_TEMPLATE)
    print('Loading template file ', template)
    name = request.args.get('name', None)
    if name:
      print('Customizing for name ', name)
    company = request.args.get('company', None)
    if company:
      print('Customizing for company ', company)
    resume_html = return_resume(template, name, company)
    return resume_html


# This is only used when running locally. When running live,
# gunicorn runs the application.
if __name__ == "__main__":
    app.run(debug=True, host="0.0.0.0", port=int(os.environ.
get("PORT", 8080)))
```

Now that all five files are copied together, `.dockerignore`, `Dockerfile`, `main.py`, `requirements.txt`, and `english.html`, we can just run the following command to deploy our container:

```
gcloud run deploy
```

You should see an output similar to the following (I just included an excerpt) and, again, you may be asked to authenticate or enable certain APIs during the first run:

```
Deploying from source. To deploy a container use [--image]. See
https://cloud.google.com/run/docs/deploying-source-code for
more details.

...

Please specify a region:

...

[27] us-central1

...Please enter numeric choice or text value (must exactly
match list item):  27

...

Building using Dockerfile and deploying container to Cloud Run
service [cloudrun] in project [cloud-developers-365616] region
```

```
[us-central1]
OK Building and deploying...
...Service [cloudrun] revision [cloudrun-00005-tux] has been
deployed and is serving 100 percent of traffic.
Service URL: https://cloudrun-4slsbxpeoa-uc.a.run.app
```

As you will see, there are a few questions asked during the deployment:

- The first question is the location of the source code, the current directory being the default.
- Then, the service name needs to be filled in. It will be prepopulated with a normalized version of the current directory name.
- At this point, we should choose a region for our deployment using a numerical code.

Once we have answered all three questions, the actual deployment will begin, and the service URL will be displayed in the last line of the output. As usual, we can use it to test that everything works as expected, and add our standard parameters too, with a sample URL like this one:

```
https://cloudrun-4slsbxpeoa-uc.a.run.app/?template=english.
html&name=John+Smith&company=StarTalent
```

Our application is ready and running, but we can still debug it if we detect any issues and want to identify its root cause.

Debugging in Cloud Run

Fortunately, in the case of Cloud Run, we can use Cloud Code to leverage **Skaffold Debug** (https://skaffold.dev/docs/workflows/debug) and debug our containers on an emulator. Debugging requires a Cloud Code-ready Cloud Run application that includes a skaffold.yaml configuration file and a launch.json file of type cloudcode.cloudrun.

This option is compatible with all the IDEs that Cloud Code supports.

You can read more about this process on the following page of the official documentation: https://cloud.google.com/code/docs/shell/debug-service.

How much does it cost to run your code on Cloud Run?

Cloud Run supports two different pricing models (https://cloud.google.com/run/pricing), which we should get familiarized with because it may affect the way in which we develop our components to reduce the global cost of our applications or services:

- **Request-based**: In this model, we don't pay for idle periods since no CPU is allocated when there is no traffic, but a fee will be charged for each request that is sent to our container. This model works better for services with a small number of requests or those where requests happen at specific times of the day.

- **Instance-based**: In this model, we pay a fixed fee for the entire lifetime of the container instance and CPU resources are permanently allocated. In this model, there are no added per-request fees. This model will be interesting for services expecting intensive traffic, with a considerable number of requests happening consistently during the day. In these cases, the cost of an instance will be significantly lower than the total cost of the request-based pricing model.

As we discussed in the case of App Engine, these two models are not mutually exclusive but complement each other, so we can use the request-based model for microservices or components with a low volume of requests, and we can choose an instance-based model for those that will receive traffic constantly.

Tips and tricks for running your code on Cloud Run

Earlier in the chapter, when I started introducing Cloud Run, I mentioned the need to be prepared for imminent disposal. If we want to receive a warning when Cloud Run is about to shut down one of our container instances, our application can trap the SIGTERM signal. This enables our code to flush local buffers and persist local data to an external data store. To persist files permanently, we can either integrate with Cloud Storage or mount a **network filesystem** (**NFS**).

You can find an example of how to handle the SIGTERM signal here: `https://cloud.google.com/run/docs/samples/cloudrun-sigterm-handler`

Cloud Run is also an interesting option for implementing **Continuous Delivery** and **Continuous Deployment**. If you store your source code in GitHub, Bitbucket, or Cloud Source Repositories, you can configure Cloud Run to automatically deploy new commits using a **Cloud Build trigger**. When we use a Cloud Build trigger to build containers, the source repository information is displayed in the Google Cloud console for our service after we deploy it to Cloud Run.

You can read more about this topic in the following section of the documentation: `https://cloud.google.com/run/docs/continuous-deployment-with-cloud-build`

Regarding networking, a Cloud Run service can either be reachable from the internet, or we can restrict access in three ways:

- Specify an access policy using Cloud IAM. You can read more about this topic and its implementation here: `https://cloud.google.com/run/docs/securing/managing-access`.

- Use ingress settings to restrict network access. This is useful if we want to allow only internal traffic from the **Virtual Private Cloud** (**VPC**) and internal services. You can find more information here: `https://cloud.google.com/run/docs/securing/ingress`.

- Only allow authenticated users by using Cloud **Identity-Aware Proxy** (**IAP**). Read more on how to enable it in the following section of the documentation: `https://cloud.google.com/iap/docs/enabling-cloud-run`.

Continuing with this topic, Cloud Run container instances can reach resources in the VPC network through the serverless VPC access connector. This is how our service can connect with Compute Engine virtual machines or products based on Compute Engine, such as Google Kubernetes Engine or Memorystore. You can read more about this topic on this page of the documentation: `https://cloud.google.com/run/docs/configuring/connecting-vpc`.

Regarding load balancing and disaster recovery, data and traffic are automatically load balanced across zones within a region. Container instances are automatically scaled to handle incoming traffic and are load balanced across zones as necessary. Each zone maintains a scheduler that provides this autoscaling per zone. It's also aware of the load other zones are receiving and will provide extra capacity in-zone to make up for any zonal failures. You can find more information about this topic in this section of the documentation: `https://cloud.google.com/architecture/disaster-recovery#cloud-run`.

Directly connected to the previous topic, having multiple instances located across different zones may be a problem unless you implement **data synchronization** to ensure clients connecting to a Cloud Run service receive a uniform response, for example, if you are using WebSockets to implement a chat server. For this purpose, you will need to integrate an external data storage system, such as a database or a message queue. In scenarios like these, it will also be important to implement **session affinity**, so that clients stay connected to the same container on Cloud Run throughout the lifespan of their connection. External storage is again the solution to mitigate both problems, as you can read in this section of the documentation: `https://cloud.google.com/run/docs/triggering/websockets`.

If we are providing public services for users in different locations across the world, we can choose to serve traffic from multiple regions by deploying our services in these regions and setting up an external HTTP(S) load balancer. You can read more about this topic, including how to set it up, in the following section of the documentation: `https://cloud.google.com/run/docs/multiple-regions`.

Directly connected with the previous topic, we can also front a Cloud Run service with a **Content Delivery Network** (**CDN**) to serve cacheable assets from an edge location closer to clients, thus returning faster responses. Both **Firebase Hosting** and **Cloud CDN** provide this capability. You can read more about how to set up Cloud CDN for serverless compute products on the following documentation page: `https://cloud.google.com/cdn/docs/setting-up-cdn-with-serverless`.

Finally, if you are interested in understanding which services work well with Cloud Run and which are not yet supported, Google Cloud maintains an up-to-date documentation page with this information together with an introduction to **Integrations**, a new feature in preview at the time of writing, which will help us enable and configure complicated integrations directly from the UI, such as connecting to a Redis instance or mapping our custom domains to Cloud Run using an external load balancer. You can read all about these topics on the following documentation page: `https://cloud.google.com/run/docs/integrate/using-gcp-services#integrations`.

I hope you found these tips useful. With them, we finished the list of serverless products that Google Cloud offers for running our code. Now is the perfect time to summarize and compare all the offerings so that you can better choose when to use each.

Choosing the best serverless option for each use case

After reviewing Cloud Functions, App Engine, and Cloud Run, let's compare them to clarify when each fits best.

Cloud Functions is a function as a service offering, while App Engine and Cloud Run are platform as a service offerings to deploy code and containers, respectively, requiring a bit more work to prepare a deployment. While all three can overlap as alternatives for many use cases, there are some differences to consider before making our choice.

Cloud Functions is the simplest way to turn a function into a microservice and works well for event-driven scenarios. However, if you are used to working with containers, App Engine flexible and Cloud Run will probably be your favorite choices. App Engine standard would be the best choice for simpler applications that need fast scaling and can run in a limited sandbox.

App Engine can bundle multiple services within a single application, while we would need many separate Cloud Function instances to implement a complex architecture. The ability to share data among services makes App Engine a very good choice for microservice-based applications, while Cloud Functions is great for simple and single tasks.

Cloud Functions and Cloud Run are both good for hosting webhook targets. Generally, Cloud Functions is quick to set up, good for prototyping, and ideal for lower volume workflows with lightweight data as input, while Cloud Run provides more flexibility and can handle larger volumes using concurrency.

Use Cloud Run in the following cases:

- You're using languages or runtimes not supported in Cloud Functions
- You want longer request timeouts (up to 15 minutes)
- You're expecting large volumes and need concurrency (up to 80 concurrent requests per container instance)

If you need a zero-scaling solution, App Engine flexible will probably not be a good choice for you. If you want to host an always-on service, you will need to decide whether either cost or speed is more important to you, because Cloud Run is significantly cheaper, particularly in low-traffic scenarios, but App Engine is usually faster and has a lower latency.

Speaking about costs, App Engine standard includes a generous free tier, which can make it an interesting option if cost is a priority and its limitations are not a problem. Also, payment models can make a big difference between services, since sometimes a request-based fee will make sense, while in others a monthly fixed fee will help us save a lot of money.

If you need portability, all the options based on containers may be the best choice for you. If you need to run one-offs and scheduled tasks or need to complete a big volume of repetitive tasks, Cloud Run jobs and array jobs can be the best option.

Cloud Run services are great for code that handles requests or events. Example use cases include the following:

- **Websites and web applications**: Build a web app using our favorite stack, access our SQL database, and render dynamic HTML pages
- **APIs and microservices**: We can build a REST API, a GraphQL API, or private microservices that communicate over HTTP or gRPC
- **Streaming data processing**: Cloud Run services can receive messages from Pub/Sub push subscriptions and events from Eventarc

Speaking of versatility, Cloud Run has better Docker image support and more memory-CPU pairs to choose from, which may fit a lot of scenarios.

If we think about regional scope, Cloud Run allows the deployment of a service to multiple regions using a single project, which is not the case for App Engine. Besides, using Anthos, we can go beyond Google Cloud and host a part of our components in other cloud providers, making portable hybrid cloud applications easy to deploy.

Last but not least, please remember that this is not a choice to make just once, because Google Cloud serverless solutions are not mutually exclusive. by combining all of them, we can build amazing architectures where each microservice runs in the serverless platform that better fits its needs.

Summary

In this chapter, we discussed how Cloud Run works, what its basic concepts are, and what the two environments that we can choose from are. We also covered how we can use containers to run our code, how we can debug our applications and services, and how much it costs to run our code using Cloud Run.

Finally, we compared the three serverless options against each other and identified specific areas where not all of them will perform equally, so we can make the best choice, combining them to build applications that take the best of each product.

In the next chapter, we will take containerization to the next level with Google Kubernetes Engine, a very interesting alternative for container orchestration, especially when we are managing complex architectures with hundreds or thousands of containers.

Further reading

To learn more about the topics that were covered in this chapter, take a look at the following resources:

- *A Brief History of Containers From the 1970s Till Now* (`https://blog.aquasec.com/a-brief-history-of-containers-from-1970s-chroot-to-docker-2016`)

- *Microservices may be the new "premature optimization"* (`https://ptone.com/dablog/2015/07/microservices-may-be-the-new-premature-optimization/`)

6

Running Containerized Applications with Google Kubernetes Engine

In the previous chapter, we discussed different serverless options available for running our code on Google Cloud. If you remember, most of those services were internally implemented using containers.

In this chapter, we will continue focusing on the concept of containerization, but this time, we will cover how to deploy and manage applications and services that use many containers.

First, I will introduce **Google Kubernetes Engine** (**GKE**) and then discuss its key concepts and features, including topics such as security, scalability, monitoring, and cost optimization. I will then talk about the similarities and differences between GKE and Cloud Run, providing some hints on when to choose each.

If you found the examples in the previous chapter to be too lightweight, I have good news for you: a good part of this chapter will be dedicated to a detailed hands-on example. I intend to gradually increase the presence and the scope of these examples as we progress through this book.

We'll cover the following main topics in this chapter:

- Introducing Google Kubernetes Engine
- Deep diving into GKE – key concepts and best practices
- Comparing GKE and Cloud Run – when to use which
- GKE hands-on example

Introducing Google Kubernetes Engine

Google Kubernetes Engine (**GKE**) is a Google Cloud service that provides a managed environment where we can deploy, manage, and scale our containerized applications.

A GKE cluster is formed by multiple Compute Engine instances and consists of at least one **control plane** and multiple worker machines called **nodes**. All of them run the **Kubernetes** (`https://kubernetes.io`) cluster orchestration system.

GKE (`https://cloud.google.com/kubernetes-engine`) works with containerized applications. As we mentioned in the previous chapter, these are applications packaged into platform-independent, isolated user space instances – for example, by using Docker (`https://www.docker.com`). Containerized applications are also referred to as **workloads**, which are packaged into a container before they can be used to run an application or host a set of batch jobs.

Kubernetes also provides different mechanisms to help us interact with our cluster. We can use Kubernetes commands and resources to deploy and manage our applications, perform administration tasks, set policies, and monitor the health of our deployed workloads.

Deep diving into GKE – key concepts and best practices

Starting with basic concepts, a **Kubernetes Pod** (`https://cloud.google.com/kubernetes-engine/docs/concepts/pod`) is a self-contained and isolated logical host that contains all the needs, at a systemic level, of the application it will serve. Pods are the smallest deployable objects in Kubernetes, representing a single instance of a running process in our cluster and can host one or more containers. All the containers within the same Pod share their resources and are managed as a single entity.

Each Pod has a unique IP address and all the containers in the Pod share the same IP address and network ports, using localhost to communicate with each other. Shared storage volumes may also be present and shared among the containers.

A **Kubernetes Service** (`https://kubernetes.io/docs/concepts/services-networking/service/`) is an abstract way to expose an application running on a set of pods as a network service. Kubernetes will assign each Pod a unique IP address, and each set of pods will get a single DNS name, and they will be used to load-balance across them.

Let's take a look at the architecture of a GKE cluster.

GKE cluster architecture

In the cluster architecture of GKE (`https://cloud.google.com/kubernetes-engine/docs/concepts/cluster-architecture`), the **control plane** runs processes that provide features such as the Kubernetes API server (which makes it our endpoint for management), scheduler, and core resource controllers:

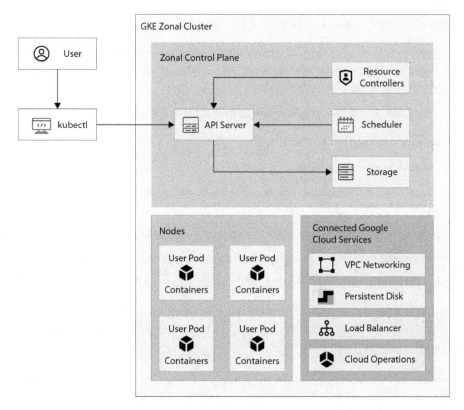

Figure 6.1 – Architecture of a zonal cluster in GKE

We interact with the cluster through Kubernetes API calls, which can be made using HTTP or gRPC, indirectly using the Kubernetes command-line client (**kubectl**) to run commands, or using the Google Cloud console.

Nodes are the worker machines that run our containerized applications and other workloads. These are individual **Compute Engine VM instances** that will be created by GKE for our cluster.

A **node** also runs those services required by the containers within our workloads, including the runtime and the **Kubernetes node agent** (*kubelet*), which communicates with the control plane and will take care of starting and running our containers as scheduled.

The control plane controls all nodes and receives periodical status updates from each of them. As we will discuss shortly, we can decide to manually check and decide on the life cycle of each node, or we can let GKE repair and upgrade our cluster's nodes automatically.

A few other special containers are also part of GKE and run as per-node agents to provide functionality such as **log collection** and **intra-cluster network connectivity**.

Advanced cluster management features

Kubernetes can provide many of the features we already discussed for other managed services in Google Cloud, including automatic management, monitoring and liveness probes for application containers, automatic scaling, and rolling updates.

There are also some advanced cluster management features that Google Cloud provides:

- Load balancing to help us distribute traffic

- Node pools to improve flexibility

- Automatic scaling and node auto-provisioning, while optionally letting us define the minimum and maximum number of nodes

- Automatic upgrades of the node software to keep it up to date with the cluster control plane version

- Node auto-repair to maintain node health and availability, with periodic health checks

- Logging and monitoring with Google Cloud's operations suite for visibility

All these features can be used in two different ways. Let's explore their similarities and differences.

GKE operation modes

We can choose to run GKE using two different operation modes (`https://cloud.google.com/kubernetes-engine/docs/concepts/types-of-clusters`), with the best choice depending on how much flexibility, responsibility, and control we want to have over our clusters:

- **Autopilot**: As its name suggests, it manages both cluster and node infrastructure for us so that we can focus on our workloads and only pay for the resources used to run our applications.

- **Standard**: In this mode, we can decide which configuration to use for our production workloads, including the number of nodes that will be used and paid for. This means more flexibility and control at the cost of more responsibility and work.

The software running on GKE cluster control planes is automatically upgraded, so we can enjoy new features and security fixes as soon as they are available. New features are listed as Alpha, Beta, or Stable, depending on their status. GKE will include the last two in newer builds, but Kubernetes Alpha features will only be available in special GKE Alpha clusters that still can be requested and used if we want to (`https://cloud.google.com/kubernetes-engine/docs/concepts/alpha-clusters`).

At the other end of the life cycle, GKE also handles any potential feature deprecations so that they have the minimum possible impact on our running applications and services (`https://cloud.google.com/kubernetes-engine/docs/deprecations`).

Cluster types based on availability

Another classification of GKE clusters, apart from their mode of operation, can be done using availability as a criterion (`https://cloud.google.com/kubernetes-engine/docs/concepts/types-of-clusters`).

While all clusters created in autopilot mode are regional, cluster types in GKE standard mode can be either zonal (single-zone or multi-zonal) or regional.

> **Important**
>
> Once a cluster has been created, we won't be able to change it from zonal to regional, or from regional to zonal. Instead, we will have to create a new cluster and then migrate traffic from the old cluster to the new one.

Zonal clusters have a single control plane running on a single zone, but we can choose to run our workloads in one or multiple zones, depending on our availability requirements:

- A **single-zone cluster** uses a single zone to run both its workloads and a single control plane that manages them. Due to this limitation, if an outage affects this zone, there will be downtime for all our workloads.

- A **multi-zonal cluster** has nodes in different zones, but its control plane will be replicated in a single zone. This means that our workloads would still run during an outage affecting a single zone, even if it is the zone running the control plane, but in that case, we would lose management capabilities until the service is restored in that zone.

Regional clusters run workloads and the control plane in the same region, but here, the control plane replicates multiple times in different zones, which makes it much more resistant to any potential outages. Nodes can run on either a single zone or multiple ones, but they must be located within the same region used by the control plane.

By default, GKE creates three replicas of each node in different zones, but we can optionally customize this number and the regions where they will be run ourselves when we create a cluster or add a new node pool.

> **Important**
>
> Regional clusters are recommended for production workloads clusters because they offer higher availability compared to zonal clusters.

Node pools and node taints for easier management

Since the number of containers in GKE can grow into the thousands, there are some features and concepts that can make it easier for us to manage big environments.

Node pools (`https://cloud.google.com/kubernetes-engine/docs/concepts/node-pools`) are groups of nodes within a cluster that all have the same configuration. To assign a node to a pool, we must use a **Kubernetes node label**, `cloud.google.com/gke-nodepool`, which will have the node pool's name as its value.

Nodes in a node pool can be created, upgraded, and deleted individually, but they share the same configuration and any changes in this configuration will be applied globally to all the nodes in the pool. Node pools can also be resized by adding or removing nodes.

If we create a new node pool in multi-zonal or regional clusters, it will be automatically replicated to all other zones. Similarly, deleting a node pool from a zone in these cluster types will remove it from the other zones as well.

> **Note**
>
> The multiplicative effect of multi-zonal or regional clusters may increase the costs of creating node pools. Take this into account when defining your architecture.

When a new workload is submitted to a cluster, it will include custom requirements for CPU, memory, and additional resources that the new Pod requires. The scheduler will read and evaluate those requirements and automatically select in which node (that can satisfy such requirements) it will be run.

Sometimes, we may need to have more control over this otherwise automatic node selection process. In these cases, we have two different tools that can be used together to ensure that certain pods are not assigned to unsuitable nodes:

- **Node taints** can be used to label nodes to let the scheduler know that specific pods should be avoided in those nodes
- **Tolerations** can be used to indicate which specific pods can be run on a tainted node

Taints are defined as the combination of a key-value pair and are associated with one of the following effects:

- *NoSchedule*: Only pods tolerating this taint will be scheduled on the node. Formerly existing pods won't be affected when a new taint is defined, so they won't be evicted from the node even if they don't tolerate the taint.

- *PreferNoSchedule*: Only pods tolerating this taint will be scheduled on the node, unless there is no other left, making this a more flexible limitation.

- *NoExecute*: This is the hardest limitation, where only pods tolerating this taint will be scheduled on the node, but even those pods that were running before the taint was created will be evicted from the node. While there are different options to add node taints to our clusters, using GKE has some interesting benefits that justify its use:

  - Taints will be preserved whenever we need to restart a node, even if we decide to replace it

  - If we add new nodes to a node pool or clusters, taints will be automatically created for each of them

  - If new nodes are added to our cluster due to autoscaling, the corresponding taints will be automatically created, too

Let's explain how taints and tolerations work with an example.

We can use the following command to create a node pool on a cluster that has a taint with a key value of `dedicated=testing` combined with a `NoSchedule` effect:

```
gcloud container node-pools create example-pool \
--cluster example-cluster \
--node-taints dedicated=testing:NoSchedule
```

This means that, by default, the scheduler will prevent pods that contain a key value of `dedicated=testing` in their specification from being scheduled on any of these nodes.

However, we can configure specific pods to tolerate a taint by including the `tolerations` field in the pods' specification. Here's part of a sample specification:

```
tolerations:
- key: dedicated
  operator: Equal
  value: testing
  effect: NoSchedule
```

The Pod using this specification can be scheduled on a node that has the `dedicated=testing:NoSchedule` taint because it tolerates that specific taint; otherwise, the `NoSchedule` effect would prevent it from being scheduled on that node.

You can read more about node taints in the following section of the official documentation: `https://cloud.google.com/kubernetes-engine/docs/how-to/node-taints`.

Best practices for cost efficiency in GKE

Spot VMs (`https://cloud.google.com/kubernetes-engine/docs/concepts/spot-vms`) are a very interesting concept for GKE since they can help us save a lot of money.

The concept is very simple: Spot VMs are virtual machines offered at a significantly lower price because they can be required back at any time. If that happens, we will receive a termination notice and the VM will be gone in 30 seconds.

We can use Spot VMs in our clusters and node pools to run stateless, batch, or fault-tolerant workloads that can tolerate disruptions caused by their ephemeral nature. Besides, and in contrast to **Preemptible VMs** (`https://cloud.google.com/kubernetes-engine/docs/how-to/preemptible-vms`), which expire after 24 hours, Spot VMs have no specific expiration time and will only be terminated when Compute Engine needs the resources elsewhere. Spot VMs are also compatible with **cluster autoscaler** and **node auto-provisioning**.

It is also good to use a node taint to ensure that GKE does not schedule critical workflows or standard ones that take longer than 30 seconds to run, to avoid early terminations. For this reason, a best practice is to ensure that our cluster always includes at least one node pool using standard Compute Engine VMs; otherwise, they won't be able to run anywhere and that would be a serious flaw in our architecture.

And if, once we are in production, we are planning to keep running our application or service in Google Cloud for quite a long time, we may want to consider signing up for **committed use discounts**, where we can get up to a 70% discount on VM prices in return for committing to paying for either 1 or 3 years of use. You can learn more about these discounts in the following section of the documentation site: `https://cloud.google.com/compute/docs/instances/signing-up-committed-use-discounts`.

Next, we'll discuss how to implement three of the most important elements of any application and service: storage, networking, and security.

Storage in GKE

Storage on GKE (`https://cloud.google.com/kubernetes-engine/docs/concepts/storage-overview`) can be implemented either using Kubernetes storage abstractions, including ephemeral and persistent volumes, or using a managed storage product offered by Google Cloud, such as a database (Cloud SQL, Cloud Spanner, and so on), **Network Attached Storage** (**NAS**), which can be implemented using FileStore, or block storage using persistent disks.

Networking in GKE

Speaking about networking in GKE (`https://cloud.google.com/kubernetes-engine/docs/concepts/network-overview`) requires a change of focus to considering how pods, services, and external clients communicate rather than thinking about how our hosts or **virtual machines** (**VMs**) are connected.

IP addresses are key for the Kubernetes networking model, with each GKE service having a stable IP address during its lifetime, often also referred to as a **ClusterIP**, while the IP addresses of pods are ephemeral.

Communication is possible thanks to a few components working together. First, management is made possible thanks to a set of firewall and allow egress rules that are automatically created for `*.googleapis.com` (Google Cloud APIs), `*.gcr.io` (Container Registry), and the control plane IP address.

Second, a component called **kube-proxy** is watching the Kubernetes API server, adding and removing destination NAT rules to the node's iptables to map the ClusterIP to healthy pods. When traffic is sent to a Service's ClusterIP, the node selects a Pod at random and routes the traffic to that Pod.

Internal and external name resolution services are offered using a combination of a component called **kube-dns** and Google's own **Cloud DNS**. Here, load balancing is offered in three different flavors: external, internal, and HTTPS. This helps provide connectivity in all possible scenarios.

Security in GKE

Security is very important in container-based architectures. A **layered security approach** works especially well in these scenarios because protecting workloads means protecting the many layers of the stack, including the contents of our container image, the container runtime, the cluster network, and access to the cluster API server. Combining this approach with the **principle of least privilege**, which means that a user should be given the most restricted set of privileges needed to complete their task, can be a very good practice, too.

Authentication in Kubernetes is provided using two types of accounts: user and service. It's important to understand the difference between a **Kubernetes service account** and a **Google Cloud service account**. The first is created and managed by Kubernetes but can only be used by Kubernetes-created entities, such as pods, while the second is a subtype of a Kubernetes user account.

To implement the principle of least privilege, we should try to use Kubernetes service accounts whenever possible, since their scope will be limited to the cluster where they were defined. This is as opposed to Google Cloud service accounts, which have a broader scope and may have too many permissions granted.

Finally, **audit logging** (`https://cloud.google.com/logging/docs/audit`) can be used to centralize all events that occur in GKE environments. Logged information can be used for forensic analysis, real-time alerting, or usage pattern detection.

Deploying applications on GKE

If we are planning to deploy a workload on GKE, we should make sure to complete the tasks mentioned in the following checklist before proceeding:

- Select a **node image** (`https://cloud.google.com/kubernetes-engine/docs/concepts/node-images`) for each container

- Choose the operation mode and level of availability for each cluster

- Define the resources to be allocated to each container (`https://kubernetes.io/docs/concepts/configuration/manage-resources-containers/`)

- Decide which storage types we will be using, if any

- Configure networking, if needed

- Make sure to follow the best practices for security, especially regarding service accounts and credential storage

- Use node taints, tolerations, and node pools (labeling) if we will be managing a considerable number of nodes

- Review the architecture to identify potential opportunities for cost optimization

Once we have made our choices, it's time to deploy and manage our containerized applications and other workloads on our Google GKE cluster. For this purpose, we will use the Kubernetes system to create Kubernetes controller objects, representing the applications, daemons, and batch jobs running on our clusters.

We can create these controller objects through the Kubernetes API or using **kubectl**, a command-line interface to Kubernetes installed by `gcloud` that allows us to create, manage, and delete objects.

Kubernetes controller objects are defined using **YAML configuration files** that contain the desired configuration values for one or more properties of the object.

These YAML files can then be passed to either the Kubernetes API or **kubectl**, the Kubernetes command-line tool, which will apply the requested operations to the referenced object.

Kubernetes provides different controller object types that allow us to create their associated workloads.

Some common types of workloads are as follows:

- **Stateless applications**: These are applications that do not preserve their state and save no data to persistent storage. They can be created using a **Kubernetes Deployment**.

- **Stateful applications**: These are applications that use persistent storage because they require their state to be saved or persistent. They can be created using a **Kubernetes StatefulSet**.

- **Batch jobs**: These represent finite, independent, and often parallel tasks that run to their completion, such as automatic or scheduled tasks. They can be created using a **Kubernetes Job**.

- **Daemons**: These perform ongoing background tasks in their assigned nodes without the need for user intervention. They can be created using a **Kubernetes DaemonSet**.

Google Cloud provides continuous integration and continuous delivery tools to help us build and serve application containers. We can use **Cloud Build** (https://cloud.google.com/build) to build container images (such as Docker) from a variety of source code repositories, and **Artifact Registry** (https://cloud.google.com/artifact-registry) or **Container Registry** (https://cloud.google.com/container-registry) to store and serve our container images.

We can use two methods to make changes to our objects: imperative commands and declarative object configuration:

- **Imperative commands** are the traditional command-line statements that we can use for one-off tasks where we want to quickly create, view, update, or delete objects using kubectl, without the need to use configuration files or even have a deep knowledge of the object schema. An example would be a command to create a new cluster or a command to change a single property of an object.

- **Declarative object configuration** uses a text-based configuration file that contains the values for each of the parameters. This option does not require us to specify specific commands to make changes because we will be passing the full configuration file. In this case, kubectl will read the live object and compare the value of each of its properties with the ones included in the provided configuration file. If any changes are required, they will be applied by sending one or more patch requests to the API server.

You can read more about deploying workloads on GKE in the following section of the official documentation: https://cloud.google.com/kubernetes-engine/docs/how-to/deploying-workloads-overview.

Next, let's discuss how we can make our application scale up and down with GKE.

Scaling an app in GKE

Scaling an application (https://cloud.google.com/kubernetes-engine/docs/how-to/scaling-apps) means increasing or decreasing its number of replicas, depending on the level of traffic and demand.

We can use the kubectl autoscale command to make changes manually, but this will force us to continuously watch specific metrics to find out when changes are needed. The alternative is to let the autoscaler do this for us.

For example, we can do this with a command like this:

```
kubectl autoscale deployment my-app \
--max 6 --min 1 --cpu-percent 40
```

The CPU utilization will be used to scale our application, where new replicas will be added if the CPU stays above 40% for some time. Notice how we also define the minimum and maximum number of replicas, set to 1 and 6, respectively.

While there are a few predefined metrics to use for autoscaling, we can also use our own, too. You can read how to do it in this tutorial: https://cloud.google.com/kubernetes-engine/docs/tutorials/custom-metrics-autoscaling.

While we recently mentioned that regional clusters are a better option for high availability scenarios, this happens at the expense of a slower propagation of configuration changes, which may affect the speed at which they can scale up, so all the pros and cons need to be considered before deciding what cluster type to use.

In the specific case of multi-zonal node pools, for example, we may suffer higher latency, more costly egress traffic between zones, and the potential lack of some features in some zones, such as GPUs, which means that we will also need to study each case before deciding which specific zones to use.

You can find some more interesting tips and best practices for increasing both availability and performance in our workloads at https://cloud.google.com/kubernetes-engine/docs/best-practices/scalability, together with details about the different quotas and limits that should be considered when scaling our applications. I didn't include the specifics here since these numbers are constantly being updated.

As a final note, don't forget that scaling up is not the only way to have a more powerful cluster. Let's see how we can also achieve this using GPUs.

GPUs may be of help, too

In some specific use cases, we may prefer to have more processing power rather than a bigger number of nodes. In these scenarios, using a **graphics processing unit** (**GPU**) can be an interesting choice. In GKE autopilot and standard, we can attach GPU hardware to nodes in our clusters, and then allocate GPU resources to containerized workloads running on those nodes.

We can use these accelerators to increase the performance of resource-intensive tasks, such as **large-scale data processing** or **machine learning** (**ML**) inference and training. We can even configure multiple containers to share a single physical GPU and optimize costs (https://cloud.google.com/kubernetes-engine/docs/concepts/timesharing-gpus).

You can read more about GPUs and how to use them in this section of the official documentation: https://cloud.google.com/kubernetes-engine/docs/concepts/gpus.

Monitoring GKE applications

Logging and monitoring are key to ensuring the successful execution of our containerized workloads, and fortunately, Google Cloud has a product called **Cloud Operations for GKE** (`https://cloud.google.com/stackdriver/docs/solutions/gke`) that can help us monitor GKE clusters by combining monitoring and logging capabilities.

Using the customized Cloud Operations dashboard for GKE clusters, we can do the following:

- Display the cluster's key metrics, such as CPU use, memory utilization, and the number of open incidents
- View clusters by their infrastructure, workloads, or services
- Inspect namespaces, nodes, workloads, services, pods, and containers
- Check the health status of the Kubernetes control plane
- View GKE logs for Kubernetes clusters, node pools, pods, and containers

We have been talking a lot about GKE in this chapter, but we have missed a very important topic: how much does it cost to run an application in GKE?

Price model for GKE

Estimating the costs of running an application on GKE is not easy since architectures tend to be complex and make use of additional Google Cloud services. Let's cover some of the basics to make this process easier.

First of all, let me share some good news: GKE has a free tier that provides enough monthly credits to run a single zonal or autopilot cluster for free.

The pricing model, however, will vary depending on the mode.

For autopilot clusters, there is a flat rate of $0.10 per hour at the time of writing. The final figure should also consider the cost of additional resources used by our pods, such as CPU, memory, or ephemeral storage.

Standard mode clusters use Compute Engine instances as worker nodes, so standard Compute Engine prices are applied in this case. These clusters can also benefit from committed use agreements, which we mentioned earlier in the book, with discounts of up to 70% off.

There is also an additional cluster management fee of $0.10 per cluster per hour to consider, regardless of the cluster mode, size, or topology. On the other hand, system pods, operating system overhead, unallocated space, or unscheduled pods do not accrue any costs.

It's also important to know that GKE includes a financially-backed **service-level agreement** (**SLA**) guaranteeing an availability rate of 99.95% for the control plane of regional clusters, and 99.5% for the control plane of zonal clusters. Since the SLA is financially backed, if these availability levels are not met, customers would be compensated.

The full documentation about GKE pricing, including a useful pricing calculator, can be found in this section of the official documentation site: `https://cloud.google.com/kubernetes-engine/pricing`.

As this section comes to an end, let's discuss when we should use GKE and Cloud Run. Then, we'll be ready to start working on the hands-on examples.

Comparing GKE and Cloud Run – when to use which

In the previous chapter, we compared different serverless products while trying to understand which one was the best choice, depending on our use case. Following a similar process to compare GKE and Cloud Run, we will start by taking a look at their summarized features:

- With GKE, we have complete control over every aspect of container orchestration, from networking to storage, and stateful use cases are supported

- Cloud Run can deploy containers to production with a single command, supports any programming language, and uses Docker images, but only for stateless apps

As a small spoiler for the next chapter, Cloud Run is offered not only as a managed service but also as **Cloud Run for Anthos** (`https://cloud.google.com/anthos/run`). In this second format, Cloud Run is compatible with custom machine types, has additional networking options, and we can use GPUs to enhance our services. And the best part is that we can easily change our mind at any time later, switching from managed Cloud Run to Cloud Run for Anthos or vice versa, without having to reimplement our service.

On the other hand, if we want to run a stateful application or if we are planning to deploy a complex architecture with hundreds of containers, GKE will probably be a better choice due to its additional features that will help us manage our fleet much more comfortably. Cloud Run supports up to 1,000 containers at the time of writing this chapter, but this limit can be raised by request (`https://cloud.google.com/run/quotas`). Also, at the time of writing, GKE can handle a maximum of 100 clusters per zone, plus 100 regional clusters per region, with up to 15,000 nodes per GKE standard cluster or 1,000 per autopilot cluster.

So, long story short, if we are looking for simplicity or don't need the extra features, configuration options, and control provided by GKE, Cloud Run can be a great choice for an application running on top of stateless containers. An intermediate option would be GKE in autopilot mode. GKE standard would be the best choice for complex architecture requiring more flexibility and control.

Please remember that these options are not mutually exclusive. We can (and should) combine them to build architectures where each component is run using the most suitable combination of product and option, smartly balancing cost and complexity.

And now (drum roll!), it's time for a hands-on example that will put many of the concepts introduced in this chapter into practice.

GKE hands-on example

In this hands-on example, we will take a simple web application written in Python that implements a phonebook, using a MySQL database to store its data. Contact entries include a name and a phone number, and the application will let us view our contact list, add new contacts, and delete any of them.

We will create a Cloud SQL instance to replace MySQL and containerize the application to serve requests from a frontend running on an autopilot GKE cluster:

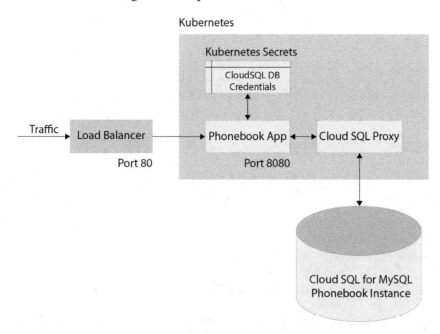

Figure 6.2 – Architecture diagram for our GKE hands-on example

First, select the project that you will use for this example by going to `https://console.cloud.google.com/projectselector/kubernetes`. If you haven't done so yet, the selector will take you to the Kubernetes API screen so that you can enable it by clicking on the blue button with the text **ENABLE**. Please be patient since this process may take a few seconds or even minutes. However, once it's completed, you will see the Kubernetes cluster page in the Google Cloud console. Please also make sure that billing is enabled for your cloud project or that you have free trial credits available.

This example uses the following billable components of Google Cloud: GKE and Cloud SQL.

> **Note**
>
> Once you have completed this example, you can avoid continued billing by deleting the resources you created during its execution. If you have the chance, run this exercise on a new project so that you can shut it down at the end. For more information, please read the *Cleaning up* section at the end of this tutorial.

Open Cloud Shell by clicking on its icon on the right-hand side of the top bar. Please keep the shell open during all the exercises since we will use environment variables to make the deployment easier. These will be lost if you close it. I recommend opening it in a new window so that you can have a more comfortable setup in full screen.

You should start by running the following commands to configure your project ID and store it in an environment variable:

```
export PROJECT_ID="<YOUR_PROJECT_ID>"
gcloud config set project $PROJECT_ID
```

Now, you must prepare your `gcloud` setup for a zonal cluster by choosing the zone closest to you. You can get a list of available zones and their corresponding regions using this command:

```
gcloud compute zones list
```

Then, assign both the zone and the associated region with an environment variable using the following commands. Here, replace `<COMPUTE-ZONE>` with your zone – for example, `us-west1-a` – and `<COMPUTE-REGION>` with your region – for example, `us-west1`:

```
export ZONE="<COMPUTER-ZONE>"
export REGION="<COMPUTER-REGION>"
```

Finally, run the following command to set up the zone:

```
gcloud config set compute/zone $ZONE
```

You may be requested to confirm this action, just press *Enter* or answer Y.

With that, we are ready to move on, but first, let's take a quick look at the code that we will be deploying in this exercise.

Taking a look at the code of the sample application

We will be working with a simple implementation of a phonebook. Each entry has a name and a number, which are stored in a MySQL table, and we can list all our contacts alphabetically, add a new contact, or delete a specific one.

The following portion of the code shows the routing rules for each operation:

```python
@app.route('/')
def print_phonebook_worker():
    return(print_phonebook())


@app.route('/add', methods=['POST'])
def add_entry_worker():
    new_name = request.values.get("name")
    new_number = request.values.get("number")
    if new_name and new_number:
        add_entry(new_name, new_number)
        return(html_ok("Entry was successfully added"))
    else:
        return(html_error("You must specify a name and a
number"))


@app.route('/delete', methods=['POST'])
def delete_entry_worker():
    entry_id = request.values.get("id")
    if entry_id and entry_id.isdigit():
        delete_entry(entry_id)
        return(html_ok("Entry was successfully deleted"))
    else:
        return(html_error("You must specify a valid ID"))
```

As you can see, the default URL, /, will show the whole list of contacts, while /add will be used to add a new contact, and /delete will be used to delete a specific entry upon specifying its internal ID. Parameters are always passed using **POST** so that they are not shown in the URL.

Containerizing the application with Cloud Build

To containerize the sample app, we will create a file named `Dockerfile` in our working directory that will contain the instructions for our container, including the version of Python, additional packages to install for dependencies, and the configuration of `gunicorn` to serve requests:

```
# Use the official lightweight Python image.
# https://hub.docker.com/_/python
FROM python:3.11.0-slim


# Copy local code to the container image.
ENV APP_HOME /app
WORKDIR $APP_HOME
COPY . ./


# Install production dependencies.
RUN pip install Flask gunicorn cloud-sql-python-
connector["pymysql"] SQLAlchemy


# Run the web service on container start-up.
# Here we use the gunicorn webserver, with one worker
# process and 8 threads.
# For environments with multiple CPU cores, increase the
# number of workers to be equal to the cores available.
CMD exec gunicorn --bind :$PORT --workers 1 --threads 8 app:app
```

We will also include `.dockerignore` in our working directory, a file containing filename patterns to be ignored when creating the container, similar to how `.gitignore` works for Git. This is done to ensure that local unwanted files don't affect the container build process. The following snippet is the content of this file:

```
Dockerfile
README.md
*.pyc
*.pyo
*.pyd
__pycache__
```

In this example, you will store your container in Artifact Registry and deploy it to your cluster from the registry. Run the following command to create a repository named `phonebook-repo` in the same region as your cluster, and wait for the operation to finish:

```
gcloud artifacts repositories create phonebook-repo \
--project=$PROJECT_ID \
--repository-format=docker \
--location=$REGION \
--description="Docker repository"
```

> **Note**
> Some of the following commands may request you to enable specific APIs if you run them on a new Google Cloud project. Just answer `Yes` and wait for the operation to complete; you should be ready to run the command.

Now, it's time for you to build your container image using Cloud Build, which is similar to running `docker build` and `docker push`, but in this case, the build happens on Google Cloud:

```
gcloud builds submit \
--tag ${REGION}-docker.pkg.dev/${PROJECT_ID}/phonebook-repo/
phonebook-gke .
```

The image will be stored in Artifact Registry (`https://cloud.google.com/artifact-registry/docs`) and you should see a lot of output on the console during the process, with a `STATUS: SUCCESS` message at the end.

Creating a cluster for the frontend

Now, it's time to create an autopilot GKE cluster for our frontend. Just run the following command and go for a coffee, since the creation process will take a few minutes:

```
gcloud container clusters create-auto phonebook \
--region $REGION
```

You may get an error if the container API was not enabled. If this is the case, just run the following command and try again:

```
gcloud services enable container.googleapis.com
```

Once the cluster has been created, you will see a message with a URL that will allow you to inspect its contents. Please open it and take a look at the different tabs in the UI. We will go back to that page once our deployment is completed.

You can also list and get detailed information about the cluster using the following two commands:

```
gcloud container clusters list

gcloud container clusters describe phonebook \
--region $REGION
```

This command will list the nodes in your container:

```
kubectl get nodes
```

Now that the cluster is ready, let's set up the database.

Creating a database instance

Now, let's create a Cloud SQL for MySQL instance to store the application data, using the following command, which may take a while to complete:

```
export INSTANCE_NAME=mysql-phonebook-instance

gcloud sql instances create $INSTANCE_NAME
```

You will be asked to enable the SQL Admin API if it wasn't already; the instance will be created afterward. This process may also take a few minutes to complete.

Now, let's add the instance connection name as an environment variable:

```
export INSTANCE_CONNECTION_NAME=$(gcloud sql instances describe
$INSTANCE_NAME \
--format='value(connectionName)')
```

Next, we will create a MySQL database:

```
gcloud sql databases create phonebook \
--instance ${INSTANCE_NAME}
```

Then, it's time to create a database user called appuser with a random password to authenticate to the MySQL instance when needed:

```
export CLOUD_SQL_PASSWORD=$(openssl rand -base64 18)
gcloud sql users create phonebookuser \
--host=% --instance ${INSTANCE_NAME} \
--password ${CLOUD_SQL_PASSWORD}
```

Please keep in mind that if you close your Cloud Shell session, you will lose the password. So, you may want to take note of the password just in case you don't complete this example in a single session.

You can display the password by running the following command:

```
echo $CLOUD_SQL_PASSWORD
```

If the password is correctly displayed, then we can move to the next section.

Configuring a service account and creating secrets

To let our app access the MySQL instance through a Cloud SQL proxy, we will need to create a service account:

```
export SA_NAME=cloudsql-proxy
gcloud iam service-accounts create ${SA_NAME} --display-name
${SA_NAME}
```

Let's add the service account email address as an environment variable:

```
export SA_EMAIL=$(gcloud iam service-accounts list \
--filter=displayName:$SA_NAME \
--format='value(email)')
```

Now, add the `cloudsql.client` role to your service account so that it can run queries:

```
gcloud projects add-iam-policy-binding ${PROJECT_ID} \
--role roles/cloudsql.client \
--member serviceAccount:$SA_EMAIL
```

Run the following command to create a key for the service account:

```
gcloud iam service-accounts keys create ./key.json \
--iam-account $SA_EMAIL
```

This command downloads a copy of the key in a file named `key.json` in the current directory.

Now, we will create a Kubernetes secret for the MySQL credentials. Secrets allow us to safely store variables instead of passing their values in plain text as environment variables, which would be a security risk since they would be readable to anyone with access to the cluster:

```
kubectl create secret generic cloudsql-db-credentials \
--from-literal username=phonebookuser \
--from-literal password=$CLOUD_SQL_PASSWORD
```

Finally, let's create another Kubernetes secret for the service account credentials since secrets can hold either key-value pairs or whole files that we can safely retrieve, as in this case:

```
kubectl create secret generic cloudsql-instance-credentials \
--from-file key.json
```

And we are ready to deploy the application.

Deploying the application

Our sample application has a frontend server that handles web requests. We will define cluster resources needed to run the frontend in a file called `deployment.yaml`. These resources are described as a deployment, which we use to create and update a **ReplicaSet** and its associated pods. This is the content of the deployment configuration YAML:

```yaml
apiVersion: apps/v1
kind: Deployment
metadata:
  name: phonebook-gke
spec:
  replicas: 1
  selector:
    matchLabels:
      app: phonebook
  template:
    metadata:
      labels:
        app: phonebook
    spec:
      containers:
      - name: phonebook-app
        # Replace $REGION with your Artifact Registry
        # location (e.g., us-west1).
        # Replace $PROJECT_ID with your project ID.
        image: $REGION-docker.pkg.dev/$PROJECT_ID/phonebook-
repo/phonebook-gke:latest
        # This app listens on port 8080 for web traffic by
        # default.
        ports:
```

```
          - containerPort: 8080
        env:
          - name: PORT
            value: "8080"
          - name: DB_USER
            valueFrom:
              secretKeyRef:
                name: cloudsql-db-credentials
                key: username
          - name: DB_PASS
            valueFrom:
              secretKeyRef:
                name: cloudsql-db-credentials
                key: password
      - name: cloudsql-proxy
        image: gcr.io/[...](cloud-sql-proxy:latest
        args:
          - "--structured-logs"
          - "--credentials-file=/secrets/cloudsql/key.json"
          - "$INSTANCE_CONNECTION_NAME"
        securityContext:
          runAsNonRoot: true
          allowPrivilegeEscalation: false
        volumeMounts:
          - name: cloudsql-instance-credentials
            mountPath: /secrets/cloudsql
            readOnly: true
        resources:
          requests:
          memory: "2Gi"
          cpu: "1"
      volumes:
        - name: cloudsql-instance-credentials
          secret:
            secretName: cloudsql-instance-credentials
---
```

Now, it's time to deploy the resource to the cluster.

First, let's replace the variables in the template with their actual values by using the following command:

```
cat ./deployment.yaml.template | envsubst > ./deployment.yaml
```

Then, we will apply the file to perform the actual deployment with this command:

```
kubectl apply -f deployment.yaml
```

You will see a notice about defaults being applied and a link that is interesting to visit so that you can become familiar with the default values used for autopilot GKE instances.

You can track the status of the deployment using the following command, which will continue updating the status of the pods until you stop it by pressing *Ctrl + C*:

```
kubectl get pod -l app=phonebook --watch
```

The deployment will be complete when all the pods are READY. This will take a few minutes, and will finally display a message similar to this, but with a different name:

```
NAME                      READY   STATUS    RESTARTS   AGE
phonebook-gke-65fd-qs82q  2/2     Running   0          2m11s
```

Remember that you will need to press *Ctrl + C* to exit the running command.

Exposing the service

Now, it's time to expose our phonebook application. We will be using a **Kubernetes Service** for this purpose because pods are ephemeral and, since their lifetime is limited, we should use a service address to reliably access our set of pods. Adding a load balancer will also make it possible to access the phonebook app pods from a single IP address.

The "*phonebook*" service is defined in `service.yaml`. Here are the contents of this file:

```
apiVersion: v1
kind: Service
metadata:
  name: phonebook
spec:
  type: LoadBalancer
  selector:
```

```
     app: phonebook
  ports:
  - port: 80
    targetPort: 8080
---
```

Now, it's time to create the phonebook service:

```
kubectl apply -f service.yaml
```

Finally, we should get the external IP address of the service so that we can use it from our browser. For this purpose, we will run the following command:

```
kubectl get services
```

It can take up to 60 seconds to allocate the IP address. The external IP address is listed under the EXTERNAL-IP column for the *phonebook* service and will have a first value of <pending> that will be replaced with the actual IP address after a few seconds.

You will then see an output like this, but please notice that the IP address that you will see will be different than mine. Copy it or write it down, since you will need it for the next step:

NAME	TYPE	CLUSTER-IP	EXTERNAL-IP
PORT(S)	AGE		
kubernetes	ClusterIP	10.1.128.1	<none>
443/TCP	52m		
phonebook	LoadBalancer	10.1.129.138	34.79.133.73
80:31706/TCP	91s		

At this point, our application is ready for testing!

Testing the application

In your browser, go to the following URL, replacing <external-ip-address> with the EXTERNAL_IP address of the service that exposes your phonebook instance that you just wrote down:

```
http://<external-ip-address>
```

After adding a few contacts, you should see a screen similar to this:

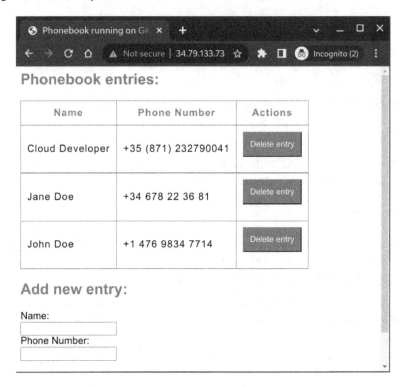

Figure 6.3 – Main screen of our phonebook running on GKE

Alternatively, you can use curl to get the HTML returned by the external IP address of the service:

```
curl <external-ip-address>
```

Next, let's take a look at how we can get more information about how our application, so we can measure its performance and perform troubleshooting.

Looking at logs and metrics

Now that our application is up and running, let's take a few minutes to add and delete contacts, and then list them. You will be doing this to generate log entries and usage metrics.

After a while, open the following address in a new tab of your web browser: https://console. cloud.google.com/kubernetes/list. Then, select the **Observability** tab. In the **Cluster** drop-down menu, select the Google Cloud project you used to run this example.

Now, you should see a screen like this one:

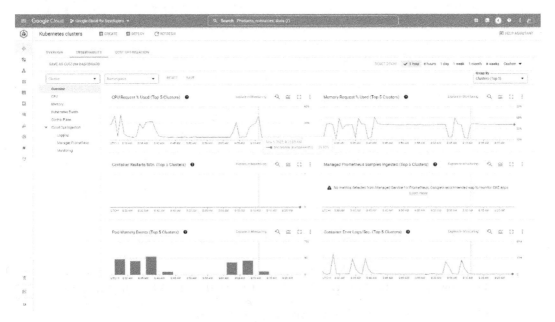

Figure 6.4 – Cloud Operations for GKE Observability tab for our example

We can see global metrics or filter by cluster and/or namespace. In the default **Observability** dashboard, we can see interesting metrics, including memory and CPU utilization, warning events, error logs per second, and restarts per minute.

In the left sidebar, we can switch to other dashboards that include specific information about areas such as memory, Kubernetes events (including statistics but also the actual events), and statistics about log ingestion and monitoring metrics.

Next to the **Observability** tab, there is a **Cost Optimization** tab, which shows how much of the requested memory and CPU our containers are using so that we can adjust these values and save money. This can also help us detect undersized clusters.

If we want to check out more details about the clusters, including the logs, we can click on the cluster's name (for example, clicking on **phonebook** in the **Overview** tab). At this point, we will be taken to a new screen that includes all kinds of information about our cluster, including configuration and features, networking, storage observability, and logs. Notice that these logs are Kubernetes logs, so any log entries generated by our Python application will not appear here.

If we want to troubleshoot our actual code running on the containers, we will need to switch to the **Workload** section using the icon menu on the left-hand side or by using the following direct link: `https://console.cloud.google.com/kubernetes/workload`.

On the new screen, once we click on the workload's name, which is `phonebook-gke` for our example, we will be able to see general and specific information about the workload, along with the revision history, events such as scaling changes, the contents of the YAML file in use, as well as the logs, which, in this case, will include events from both `gunicorn` and our Python application code. This is very useful for troubleshooting code errors.

Cleaning up

If you used a new project for this exercise, you can just shut it down to ensure that all the resources are stopped properly and that they don't cost you extra money or credits.

Otherwise, you can follow these steps to complete the cleanup:

1. Delete the GKE cluster by running the following command. The deletion process may take a few minutes to complete:

    ```
    gcloud container clusters delete phonebook \
    --region $REGION
    ```

2. Run the following command to delete the image from your Artifact Registry repository:

    ```
    gcloud artifacts docker images delete \
    $REGION-docker.pkg.dev/$PROJECT_ID/phonebook-repo/
    phonebook-gke
    ```

3. Next, delete the Artifact Registry repository with the following command:

    ```
    gcloud artifacts repositories delete phonebook-repo \
    --location=$REGION
    ```

4. Now, delete the Cloud SQL instance:

    ```
    gcloud sql instances delete $INSTANCE_NAME
    ```

5. Then, remove the role from the service account:

    ```
    gcloud projects remove-iam-policy-binding $PROJECT_ID \
    --role roles/cloudsql.client \
    --member serviceAccount:$SA_EMAIL
    ```

6. Finally, delete the service account:

    ```
    gcloud iam service-accounts delete $SA_EMAIL
    ```

What's next?

I hope this GKE example helped you understand some of the key concepts of GKE.

There is one added step that you can try to implement on your own so that your application can have a domain name and static IP address. You can read more about it here: `https://cloud.google.com/kubernetes-engine/docs/tutorials/configuring-domain-name-static-ip`.

While I chose to use commands so that you could better learn the concepts provided and see how kubectl can be used, remember that Cloud Code can help you run some tasks much more comfortably from the IDE. For example, you can manage Kubernetes clusters directly from Visual Studio Code (`https://cloud.google.com/code/docs/vscode/manage-clusters`) and also view Kubernetes task status and logs (`https://cloud.google.com/code/docs/vscode/view-logs`).

You can read more about the different capabilities of the integration with each IDE and find interesting examples, some of them about Kubernetes and GKE, on the Cloud Code extensions page: `https://cloud.google.com/code/docs`.

I have also included other interesting tutorials in the *Further reading* section so that you can see sample implementations of autopilot and standard GKE architectures using different programming languages. You can also find many more interesting examples in the Google Cloud architecture section (`https://cloud.google.com/architecture?category=containers`).

Now, it's time to review what we covered in this chapter.

Summary

This chapter started with an introduction to the basic concepts of GKE before deep diving into key topics, such as cluster and fleet management, security, monitoring, and cost optimization.

Then, we discussed the similarities and differences between GKE and Cloud Run, and when to use which.

Finally, we worked on a hands-on example to show you how to use GKE to containerize and run a web application.

In the next chapter, we will discuss how the abstraction level of containers makes them, combined with the power of Anthos, the ideal choice for hybrid and multi-cloud architectures and deployments.

Further reading

To learn more about the topics that were covered in this chapter, take a look at the following resources:

- *Google Kubernetes Engine vs Cloud Run: Which should you use?* https://cloud.google.com/blog/products/containers-kubernetes/when-to-use-google-kubernetes-engine-vs-cloud-run-for-containers

- *Kubernetes on GCP: Autopilot vs Standard GKE vs Cloud Run:* https://blog.searce.com/kubernetes-on-gcp-standard-gke-vs-cloud-run-vs-autopilot-17c4e6a7fba8

- *Choosing Between GKE and Cloud Run:* https://medium.com/@angstwad/choosing-between-gke-and-cloud-run-46f57b87035c

- *Best practices for running cost-optimized Kubernetes applications on GKE:* https://cloud.google.com/architecture/best-practices-for-running-cost-effective-kubernetes-applications-on-gke.

- *How to find – and use – your GKE logs with Cloud Logging:* https://cloud.google.com/blog/products/management-tools/finding-your-gke-logs

- *Create a guestbook with Redis and PHP:* https://cloud.google.com/kubernetes-engine/docs/tutorials/guestbook

- *Deploy WordPress on GKE with Persistent Disk and Cloud SQL:* https://cloud.google.com/kubernetes-engine/docs/tutorials/persistent-disk

- *Deploying Memcached on GKE:* https://cloud.google.com/architecture/deploying-memcached-on-kubernetes-engine

- *Tutorial: Using Memorystore for Redis as a game leaderboard:* https://cloud.google.com/architecture/using-memorystore-for-redis-as-a-leaderboard

- *Deploy a batch machine learning workload:* https://cloud.google.com/kubernetes-engine/docs/tutorials/batch-ml-workload

7

Managing the Hybrid Cloud with Anthos

The previous chapter covered different serverless options for running our code on Google Cloud. If you remember, most of those services were internally implemented using containers.

This chapter will discuss how to choose a cloud provider and what the associated risks of this process are. We will then focus on the potential benefits of using more than one provider, and what challenges this approach can bring to the table.

That will be the perfect time to introduce Anthos, a hybrid-cloud application management platform created by Google that helps us run our applications in multiple environments and with multiple providers.

The chapter will finish with a hands-on exercise, so you can try Anthos yourself and better understand how to make the most of some of its features.

We'll cover the following main topics in this chapter:

- The pitfalls of choosing a cloud provider
- Anthos, the hybrid-cloud management platform
- A hands-on example of Anthos

> **Important note**
> Please run bash setup.sh to copy third party files to the chapter folder before reading on.

The pitfalls of choosing a cloud provider

Public cloud providers have a key role in the digital transformation of any organization. However, deciding which one to use is not an easy task.

There are many factors to consider when choosing a cloud provider. I have put together a list with a few of the considerations listed in no specific order, but this list can still grow a lot depending on each specific situation and need:

- Geographical availability, which we should compare with our current and future expansion plans

- Technology and roadmaps, which should meet our requirements and be refreshed with new features and services over time

- Free tier and welcome offers, making it easier to test services and run proof of concepts at a limited cost or even for free

- Pricing is also a key factor, sometimes offering multi-year special contracts or bundles with other services

- Security and reliability, including service level objectives and agreements to guarantee uptime for our services

- Migration and support services, which can help us plan and execute our move to the cloud and help with any issues during and after the migration

- Legal and regulatory requirements that we will need to follow and that may reduce the list of providers that we can work with or the locations where our workloads can run

- Certifications and standards, from following best practices to specific adherences that may be indispensable for certain sectors, such as finance or pharma

- Compatibility or integration with the rest of the software and applications used by our organization, so we can find out whether we can benefit from any specific synergies or out-of-the-box integrations between products

Once we consider these and other important topics for our organization, we can assign a weight to each topic, calculate a score for each provider, and make our choice, taking into account the obtained results.

However, before choosing a provider, we should also consider some of the associated risks:

- **Dependency**: We may get so dependent on our provider and its technology that we will lose our freedom to migrate at will in the future. The situation will be critical as we get to use more cloud resources and as our code gets too bound to specific APIs and services.

- **Roadmap not meeting our expectations**: We may choose a provider who is not offering new features and services at the speed that we want or need. Or we may even see how other providers start offering services in areas where our provider doesn't.

- **Regulatory requirements**: This is a classic in every list of risks. Our list of candidates may have been reduced a lot due to either new or long-time active legal and regulatory constraints, and that may negatively affect our ability to innovate due, for example, to the need to store our corporate data in a specific country and/or follow a set of local regulations.

In summary, choosing a cloud provider is not easy, and there are many times and situations where we will think that we made the wrong choice and start to feel like a bird in a cage, and even worse, this will be a pay-per-use cage!

Introducing hybrid cloud computing

The ideal situation in many of these scenarios would be being able to choose the best environment and cloud provider, depending on our needs. We would only move to the cloud those workloads that will benefit from the move, while we would keep the rest on-premises.

And even when talking about moving workloads to the cloud, why not move some of these workloads to the provider where they will be cheaper and others to a different provider where they will be able to make the most out of the technology available?

This is what is called **hybrid cloud computing**, where an organization may have workloads running in private and public locations using multiple providers, often combining these workloads to architect a service or application that literally runs *on a hybrid cloud*:

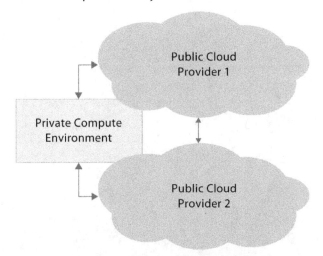

Figure 7.1 – A hybrid cloud environment

This would be an amazing opportunity from both the technical and the economic point of view, but this freedom comes at the high price of having to manage many heterogeneous components and make them work together seamlessly.

This is indeed an extremely complex scenario, but one that could also help organizations break the chains once and for all and run their workloads wherever they want. And that's what motivated Google to find a solution.

The answer is one that combines the portability of containers, thanks to its advanced level of abstraction, with a platform that supports multiple locations for deployment while providing centralized management. Let's introduce Anthos.

Anthos, the hybrid cloud management platform

In the previous chapter, we deep-dived into **Google Kubernetes Engine** (**GKE**), and you should now be familiar with its main concepts. As I mentioned at the end of that chapter, the next level in the abstraction that containers provide would be extending GKE to work in multiple environments.

This is exactly what **Anthos** (`https://cloud.google.com/anthos`) does; it provides a consistent development and operations experience whether we are using Anthos on Google Cloud, hybrid cloud, or multiple public clouds:

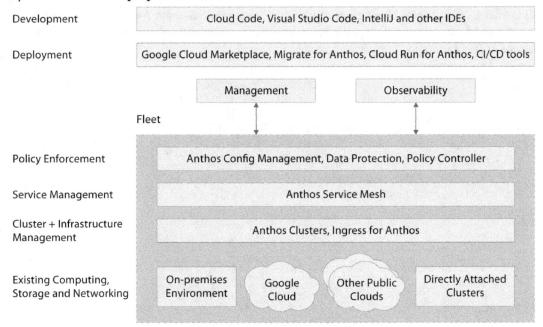

Figure 7.2 – Anthos architecture diagram

Anthos is an advanced management platform for the quick building and deployment of container-based services and applications.

Some of the key benefits of Anthos are the following:

- Code can be quickly deployed, traffic easily re-configured, and applications automatically scaled while Google takes care of all the underlying infrastructure

- Both code and configuration can be managed using Git to implement CI/CD workflows

- Observability and instrumentation are provided by Anthos Service Mesh, Cloud Monitoring, and Cloud Logging without requiring a single additional line of code

- Automatic service protection using **mutual TLS (mTLS)** and throttling (again, with no code changes needed)

- Integrated Google Cloud Marketplace allowing the quick deployment of compatible products into clusters

The main priorities for Anthos are speed, flexibility, and security, letting us manage, govern, and operate our workloads wherever we want by providing a common platform with centralized management, security, and observability.

If we need to migrate our applications and services, Anthos also includes **Migrate to Containers** (`https://cloud.google.com/velostrata/docs/anthos-migrate/anthos-migrate-benefits`), which allows us to orchestrate migrations using Kubernetes in Anthos so we can enjoy the benefits of containers. It can also *convert* legacy applications to be used in containers instead of virtual machines.

Now, let's walk through the main concepts of Anthos.

Computing environment

Anthos uses **Anthos clusters** as its primary computing environment, taking GKE beyond Google Cloud because these clusters can also run on on-prem and on other public cloud providers, allowing us to centrally manage Kubernetes installations in any of these environments.

As you will remember from the previous chapter, Kubernetes has two main parts: the *control plane* and the *node components*. These two will be hosted differently depending on where we choose to run Anthos:

- **Anthos on Google Cloud**: Google Cloud hosts the control plane, and customers can only access the Kubernetes API server. GKE manages node components in the customer's project using instances in Compute Engine.

- **Anthos on-premises**: In both Anthos clusters on VMware and Anthos clusters on bare metal, all components are hosted in the customer's on-prem data center.

- **Anthos on AWS**: All components are hosted in the customer's AWS environment.

- **Anthos on Azure**: All components are hosted in the customer's Azure environment.

While Anthos clusters are based on GKE, we can also add **conformant non-GKE Kubernetes clusters** to Anthos, taking advantage of a subset of Anthos features on our existing systems, even if we don't perform a full migration to Anthos clusters. You can read more about this topic in the following section of the official documentation site for Anthos: `https://cloud.google.com/anthos/clusters/docs/attached/how-to/attach-kubernetes-clusters`.

As you can imagine, Anthos will often be used to manage a lot of clusters. Let's introduce the concept of fleets, which can help us simplify this process a lot.

Simplified management using fleets

Another interesting concept to make management easier in Anthos is the use of fleets. A **fleet** (`https://cloud.google.com/anthos/multicluster-management/fleet-overview`) is just a logical group containing clusters and other Kubernetes resources. The benefit of a fleet is that all its elements can be viewed and managed as a group in Anthos, thus reducing the complexity when managing architectures with many components.

The most important concept for fleets is **sameness**, where Kubernetes objects sharing the same identity in different clusters are treated as the same thing. For example, services with the same namespace and service name are considered as the same service, even if they are in different clusters. This makes administration much easier because this common identity makes it possible to use a single rule for all of them instead of having to create one for each cluster. For example, we can grant the namespace *frontend* in the fleet access to the namespace *backend*, and that would make the *frontend* in each cluster able to access the *backend* in any of the other clusters in the fleet. This uplevel of management to the fleet is called **high trust** and helps narrow cluster boundaries.

A cloud project can only have one fleet associated with it, and fleet-aware resources can only be members of a single fleet at any given time. This is done to ensure that there is a single source of truth in every cluster.

You can read more about fleet-related concepts in this section of the official documentation: `https://cloud.google.com/anthos/fleet-management/docs/fleet-concepts`.

As we are speaking about making management easier, let's now introduce Service Mesh, another key component that helps simplify container orchestration.

Service Mesh for microservice architectures

Going back to the basics of Kubernetes, services are composed of many pods, which execute containers, and these containers, in turn, run services. In a microservice architecture, a single application may consist of numerous services, and each service may have multiple versions deployed concurrently. As you can see, there are a lot of different components that need to communicate with each other.

In legacy applications, communication between components is done using internal function calls. In a microservice architecture, service-to-service communication occurs over the network, and it can be unreliable and insecure, so services must be able to identify and deal with any potential network issues.

How should services respond to communication timeouts? Should there be retries, and if so, how many? And how long should we wait between retries? When a response is obtained, how can we know for sure that it is coming from the expected service?

A service mesh can solve these problems by using **sidecar proxies** to improve network security, reliability, and visibility. Each deployed service gets a second service attached to it (that's why they are called *sidecar*), which works as a proxy and forwards the information in real time for the service mesh to provide additional features and benefits.

Anthos has its own service mesh: **Anthos Service Mesh** (`https://cloud.google.com/anthos/service-mesh`), based on **Istio** (`https://istio.io/docs/concepts/what-is-istio/`), an open source implementation of the service mesh infrastructure layer.

These are some of the benefits that Anthos Service Mesh provides to applications running on Anthos:

- Fine-grained traffic control with rich routing rules, service metrics, and logs
- Automatic metrics, logs, and traces for all HTTP traffic within a cluster
- Service-to-service relationships mapping, including a graphical representation
- Secure service-to-service communication with authentication and authorization, including the support of mTLS authentication
- Easy A/B testing and canary rollouts
- Automatically generated dashboards to let us dig deep into our metrics and logs
- Service health metrics with **service-level objectives** (**SLOs**)

All these features are provided just by installing Anthos Service Mesh, and most of them require zero configuration, which can be helpful for big architectures with thousands of components. We will try some of these features in our hands-on exercise.

Anthos Service Mesh can always be installed in the control plane of our cluster, but if we run Anthos on Google Cloud, we will also have the option to use it as a managed service, making things even simpler.

And this is a perfect time to discuss networking in Anthos, a basic element of service-to-service communication.

Networking in Anthos

Given the potential complexity of microservice architectures, networking plays a crucial role in delivering requests to our workloads across different pods and environments.

The scenarios will vary depending on whether we run Anthos on Google Cloud, on-prem, or on another provider.

If we run Anthos on Google Cloud, **Network Load Balancing** (`https://cloud.google.com/load-balancing/docs/network`) will be used for the transport layer, and **HTTP(S) Load Balancing** (`https://cloud.google.com/load-balancing/docs/`) will be used for the application layer, with the advantage of both being managed services requiring no added configuration or provisioning. These two can be complemented with **Multi Cluster Ingress** (`https://cloud.google.com/kubernetes-engine/docs/how-to/ingress-for-anthos`), allowing us to deploy a load balancer that serves an application across multiple GKEs on Google Cloud clusters.

If we run Anthon on-prem, the options will vary depending on our environment: for clusters on VMware, we can use bundled options, such as **MetalLB** or **Seesaw**, or we can manually set up **F5 BIG-IP**, **Citrix**, or any other similar option. On bare metal, we can choose either to use a bundled transport layer load balancer during the cluster installation or to deploy an external one manually.

Running Anthos on AWS is compatible with multiple AWS load balancers, including **AWS Classic Elastic Load Balancers (ELB)**, **AWS Network Load Balancers**, and **AWS Application Load Balancers**, and with Anthos Service Mesh. In the case of Azure, we can set up a transport layer load balancer backed by an **Azure Standard Load Balancer**.

You can find detailed information about setting up each load balancing option on this page of the documentation site: `https://cloud.google.com/anthos/clusters/docs/on-prem/latest/how-to/setup-load-balance`.

Once each of our environments is ready, we can connect our on-premises, multi-cloud, attached clusters, and Google Cloud environments in diverse ways.

The easiest way to get started is by implementing a site-to-site VPN between the environments using **Cloud VPN** (`https://cloud.google.com/network-connectivity/docs/vpn`).

If you have more demanding latency and speed requirements, you may prefer to use **Dedicated Interconnect** (`https://cloud.google.com/network-connectivity/docs/how-to/choose-product#dedicated`) or **Partner Interconnect** (`https://cloud.google.com/network-connectivity/docs/how-to/choose-product#partner`), which both offer better performance at a higher cost.

Connect (`https://cloud.google.com/anthos/multicluster-management/connect`) is a basic element of Anthos that allows our clusters to be viewed and managed centrally from the Anthos **dashboard**.

For Anthos to work properly, all environments outside Google Cloud must be able to reach Google's API endpoints for *Connect*, *Cloud Monitoring*, and *Cloud Logging*. Attached clusters just need connectivity with *Connect*.

Another remarkably interesting feature of Anthos, and a great aid for simplifying management tasks, is centralized configuration management. Let's talk about it in the next section.

Centralized configuration management

As the size of our architecture on the cloud grows and becomes more complex, having a clear picture of the configuration of each of our nodes becomes a real challenge.

Configuration as Data can help manage this complexity by storing the desired state of our hybrid environment under version control. Once we want to make changes, we can just commit the updated configuration files, and they will be scanned for changes, and these will be directly applied. In Anthos, this is possible thanks to a unified declarative model that can be used with computing, networking, and even service management across clouds and data centers.

Anthos Config Management (`https://cloud.google.com/anthos-config-management`) uses a Git repository as the source of truth for configuration settings. Using Kubernetes concepts such as **namespaces**, **labels**, and **annotations** in our YAML or JSON files, we can define which configuration file must be applied to each component. A single commit will be translated into multiple kubectl commands across all clusters to apply the configuration. And rolling back the changes is as simple as reverting the change in Git, which will produce the kubectl commands required to undo the previous changes in the environment.

Anthos Config Management also includes **Policy Controller**, which can detect whether any API requests and configuration settings violate the policies and rules that we define and help us keep our clusters under control.

As we are talking about policy enforcement, discussing security on Anthos feels like a natural next step.

Securing containerized workloads

The key security features that complement Anthos Config Management include the following:

- Automatic code-free securing of microservices with in-cluster mTLS and certificate management. GKE on Google Cloud uses certificates provided by **MeshCA**.
- Built-in service protection using Anthos Service Mesh authorization and routing.

Binary Authorization is another feature that can improve the security in Anthos. Let's get to know a bit more about it in the next section.

Binary Authorization for a secure software supply chain

When we have thousands of containers running in production, the main concern of administrators is that we may easily lose control and no longer know which images are running in each of them.

Binary Authorization was created to ensure that only signed and authorized images are deployed in our environment. It uses signatures, also known as *attestations*, generated as an image passes through and prevents images that do not meet the defined policies from being deployed.

Apart from attestations, name patterns can also be used to generate allow-lists matching against the repository or the path name or by directly specifying a list of which specific images are allowed to be deployed.

Binary Authorization is currently available for Anthos on Google Cloud and in preview for Anthos on-prem.

Two other features that can complement our security strategies are *logging* and *monitoring*. Let's explain how they work in Anthos.

Consolidated logging and monitoring

Anthos integrates Cloud Logging and Cloud Monitoring for both cluster and system components, centralizing metrics and events, and includes entries from audit logs, making it easier to detect and troubleshoot issues.

If Anthos runs on Google Cloud, our workloads will have their logs additionally enriched with relevant labels, such as pod label, pod name, and the cluster name that generated them. Labeled logs will be easier to browse and filter using advanced queries.

Kubernetes Engine Monitoring (`https://cloud.google.com/monitoring/kubernetes-engine`) is the component that stores our application's critical metrics that we can use for debugging, alerting, or in post-incident analysis.

These features are available in the **user interface** (**UI**). Let's see what it looks like in Anthos.

Unified UI

The Anthos dashboard in the Google Cloud console provides a unified UI to view and manage our containerized applications.

The views of the dashboard provide different information and options to manage our resources. First, we can view the state of all our registered clusters and create new ones for our project. We can also see graphs about resource utilization, which are useful for optimizing our spending. Data provided by the Policy Controller can be used to identify areas where security and compliance can be improved. Configuration Manager helps us see the configuration state of our clusters at a glance and easily enables the component on those that haven't been set up yet.

For workloads running in GKE on Google Cloud, Anthos Service Mesh automatically uploads metrics and logs provided by the sidecar proxies, providing observability into the health and performance of our services without the need to deploy custom code to collect telemetry data or manually set up dashboards and charts. Finally, Cloud Monitoring and Cloud Logging help us troubleshoot, maintain, and optimize our applications.

The unified UI provides an excellent integration of the numerous features that Anthos provides. However, if Anthos seems too complicated for you, just remember that an *old friend* can help you make the most of Anthos while you let Google take care of all the complexities. Let's re-introduce Cloud Run.

Making hybrid cloud simple with Cloud Run for Anthos

Cloud Run for Anthos (`https://cloud.google.com/anthos/run/docs`) is a remarkably interesting option for developers since it abstracts away the complexities of the underlying Anthos platform, helping us focus on writing code for our workloads, so we can generate customer value in less time.

Instead of wasting our time authoring many YAML files and fine tuning our clusters, Cloud Run for Anthos manages how our services run, whether in the cloud or on-premises, while optimizing performance and resource usage, scaling to and from zero, and using all the features of Anthos Service Mesh.

Cloud Run for Anthos has some special benefits when compared with the standard Cloud Run, such as its ability to deploy containers, the support of custom machine types, additional networking features, or the use of GPUs to enhance our services.

At the time of writing this section, Cloud Run for Anthos is generally available for Anthos on Google Cloud and Anthos on-prem deployment options and is on the roadmap for multi-cloud and attached clusters.

And, speaking about how to make things simpler with Anthos, there is an especially useful marketplace you will love to know about.

Third-party application marketplace

Since the Kubernetes ecosystem is very dynamic, a lot of new third-party applications are regularly published or updated that could run on top of our existing clusters. We can use **Google Cloud Marketplace** to find and deploy any of these applications to our Anthos clusters, no matter where they are running. We can easily identify Anthos-compatible solutions because they will be marked with an Anthos badge.

Solutions available in the marketplace have direct integration with Google Cloud billing and are supported directly by the software vendor.

In the **Marketplace Solution Catalog** (`https://console.cloud.google.com/marketplace/browse?filter=solution-type%3Ak8s`), we can find solutions for storage, databases, CI/CD, monitoring and security, among many others. You can access the marketplace using this URL: `https://console.cloud.google.com/marketplace/browse?filter=solution-type:k8s&filter=deployment-env:anthos`.

Now that we have covered all the basics of Anthos, it's the perfect time to discuss its different usage and pricing options.

Anthos usage and pricing options

There are two pricing alternatives to choose from for the Anthos platform:

- **Pay-as-you-go pricing**: This is where we are billed for Anthos-managed clusters as we use them at the rates listed on the official Anthos pricing page (`https://cloud.google.com/anthos/pricing`). We can enable the Anthos API whenever we want to use this option.

- **Subscription pricing**: This provides a discounted price for a committed term, including all Anthos deployments, irrespective of environment, at their respective billing rates. This option can be enabled by contacting sales (`https://cloud.google.com/contact/?form=anthos`).

> **Note**
>
> New Anthos customers can try Anthos on Google Cloud for free, up to $800 worth of usage or for a maximum of 30 days, whichever comes first. This is perfect for running the Anthos examples provided in this book and some others available on the Google Cloud website. If you currently have an Anthos subscription, then this trial will not be available to you.
>
> To sign up, go to the Anthos page on the Cloud Marketplace (`https://console.cloud.google.com/marketplace/product/google/anthos.googleapis.com`) and click on **START TRIAL**. You can see your available credit in the Anthos section within the Google Cloud console.

We can make full use of the Anthos platform, including hybrid and multi-cloud features, by enabling the **Anthos API** in our project. Once it's enabled, we will be charged on an hourly basis for an amount based on the number of Anthos cluster vCPUs **under management**.

We can see the vCPU capacity, which is the number used for Anthos billing, of each of our user clusters by running the following command:

```
kubectl get nodes -o=jsonpath="{range .items[*]}{.metadata.
name}{\"\t\"} \
        {.status.capacity.cpu}{\"\n\"}{end}"
```

Logs and metrics from Anthos system components are collected in Cloud Logging and Cloud Monitoring for no added charge when we pay for the Anthos platform with a default retention period of 30 days.

If we don't need to register clusters outside Google Cloud and only require some of the Anthos features, we can pay for each of those features individually without enabling the entire platform. The pricing guide can be found at `https://cloud.google.com/anthos/pricing`.

The official deployment options page (`https://cloud.google.com/anthos/deployment-options`) contains valuable information about which Anthos features are enabled in each deployment option, a very recommendable read before choosing where you will run your workloads.

And with this, the theory is over, and it's time for some hands-on action using Anthos.

Anthos hands-on example

For this example, we will be simulating a **non-fungible token** (**NFT**) store (if you don't know what NFTs are, I added a link about them in the *Further reading* section at the end of the chapter). This will allow us to explore some of the key concepts that we have discussed in the last two chapters, including zonal clusters, sidecar proxies, canary deployments, and traffic splits. Let's take a look at the architecture and we will be ready to begin!

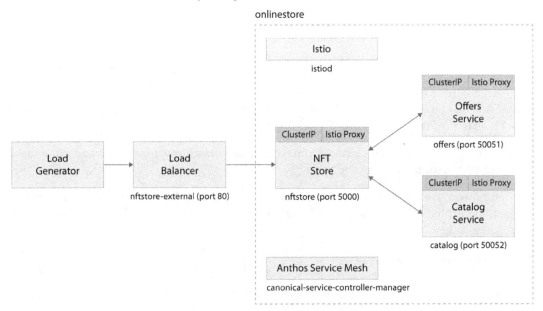

Figure 7.3 – Hand-on example architecture

First, you can choose whether you prefer to create a Google Cloud Platform project for this example or whether you prefer to reuse another. Cleaning up will be extremely easy in this example, so either of the options will work well. Whichever one you choose, please copy your project ID to the clipboard since we will need it for the next step.

Now, open Cloud Shell, set the PROJECT_ID shell variable, and check that it has the right value:

```
PROJECT_ID=<your-project-id>
echo $PROJECT_ID
```

Next, set which cloud project to use with the following command. You may be asked to authorize the configuration change with your account:

```
gcloud config set project $PROJECT_ID
```

Now, please create a directory and either copy or clone the files for this chapter from the code repository for the book. I will assume that you are running all the commands mentioned in this section from the directory for this chapter.

Then, use the following command to ensure that both GKE and Cloud Operations APIs are enabled. After running it, you may have to authenticate and authorize the command to act on your behalf to enable the APIs:

```
gcloud services enable container.googleapis.com \
--project $PROJECT_ID
```

The command may take a while to complete before confirming that the operation finished successfully. If the APIs were already enabled, you will quickly get an empty response; that's also OK.

Now we will create a GKE zonal cluster in standard mode with at least four nodes and using the e2-standard-4 machine type.

First, please choose the computing zone closest to you. You can get a list of available zones and their corresponding regions using this command; you will have to use the value of the NAME parameter for the chosen zone:

```
gcloud compute zones list
```

Once you choose a zone to use, export it as a shell variable with this command:

```
COMPUTE_ZONE="<your-GCP-zone>"
echo $COMPUTE_ZONE
```

We will also export the cluster name as a shell variable for later use. Use this one or any other you like:

```
CLUSTER_NAME="onlinestore"
```

Then, run the following command to create the cluster. The process will take a few minutes to complete. Go for a coffee if you have the chance:

```
gcloud container clusters create $CLUSTER_NAME \
    --project=$PROJECT_ID \
    --zone=$COMPUTE_ZONE \
```

```
    --machine-type=e2-standard-4 \
    --num-nodes=4
```

Let's ensure that our account has cluster-admin permissions. Just run the following command, replacing <YOUR_EMAIL> with the actual email address of the account you are using in the Google Cloud console:

```
kubectl create clusterrolebinding cluster-admin-binding \
    --clusterrole=cluster-admin \
    --user=<YOUR_EMAIL>
```

Now, we will download the latest version of the Anthos Service Mesh command-line client and make it executable using the following commands. Please note that there may be a newer version of asmcli available, and you should update the command accordingly. You can find the latest one in the *Download asmcli* section of the following documentation: https://cloud.google.com/service-mesh/docs/unified-install/install-dependent-tools:

```
curl https://storage.googleapis.com/csm-artifacts/asm/
asmcli_1.15 > asmcli
chmod +x asmcli
```

Then, it's time to install Anthos Service Mesh. We will use the --enable_all option for the installer to automatically perform any actions required to properly complete the installation, including enabling APIs, enabling workload identity, or creating namespaces and labels. Otherwise, we would have to complete all those tasks manually:

```
./asmcli install \
    --project_id $PROJECT_ID \
    --cluster_name $CLUSTER_NAME \
    --cluster_location $COMPUTE_ZONE \
    --fleet_id $PROJECT_ID \
    --output_dir ./asm_output \
    --enable_all \
    --ca mesh_ca
```

The command may take a while to run, up to 10 minutes according to its own output, depending on the features to enable, and will output a lot of information during the execution, including a warning about the Linux command **netcat (nc)** that you can safely ignore. The final line, showing a successful completion, should look the following:

```
asmcli: Successfully installed ASM.
```

Now let's get the revision name into another variable:

```
REVISION=$(kubectl -n $CLUSTER_NAME \
  get mutatingwebhookconfiguration \
  -l app=sidecar-injector \
  -o jsonpath={.items[*].metadata.labels.'istio\.io\/
rev'}'{"\n"}' | awk '{ print $1 }')
```

Use the following command to verify that the variable was assigned an actual value:

```
echo $REVISION
```

And then, let's enable automatic sidecar injection using the following command. You can safely ignore the `label not found` message that you will see after running it:

```
kubectl label namespace default istio-injection- \ istio.io/
rev=$REVISION --overwrite
```

Next, let's deploy the sample online store app to the cluster. We will apply the Kubernetes manifest to deploy our services:

```
kubectl apply -f yaml/kubernetes.yaml
```

Then, we should just wait for all the pods to be ready using the following command:

```
kubectl get pods
```

After a few minutes, you should see an output like the following, where all the pods have their two containers displayed in a `ready` status. Wait until all rows in the output show 2/2, periodically re-running the preceding command until you get the expected result. For me, it took less than one minute until I got an output like this:

```
NAME                          READY   STATUS  RESTARTS  AGE
catalog-77d69f5fbb-tf5kc      2/2     Running   0       2m57s
loadgenerator-68bf6bcb67-kg2tf 2/2    Running   0       2m57s
nftstore-fc987977f-8sx5k      2/2     Running   0       2m58s
offers-7bd86cc97-2zsjh        2/2     Running   0       2m58s
```

> **Note**
>
> Notice that each pod has two containers because the sidecar proxy is injected in all of them.

The deployment is complete at this point. Now you can find the external IP address of your Istio Gateway Ingress or service by running the following command, and you can visit the application frontend by pasting this IP address in your web browser:

```
kubectl get service nftstore-external | awk '{print $4}' | tail
-1
```

Visit http://<YOUR-EXTERNAL-IP> and play around a bit with the website, which simulates an online boutique. The front page should look like this:

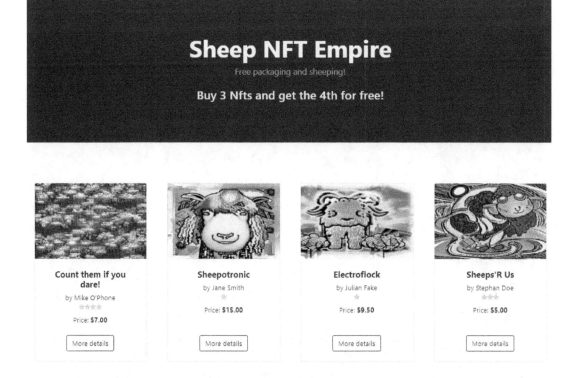

Figure 7.4 – Front page of the online boutique

You can see some products on the front page of the store, whose details are provided by the catalog service. Click on each to read more details, or you can even click on the **Add to cart** button, which, of course, is totally useless. Notice also how each time you load a web page, there is a different offer shown in yellow after the page title. These offers are provided by the offers service. You may want to look again at the architecture diagram included at the beginning of this section since all its components should make much more sense now.

After a few minutes, the load generator built-in to the online store will have generated some meaningful metrics, and your browsing will have too. At that point, it's time to look at what Anthos already knows about our service.

Search for `Anthos` in the omni-search box on the top bar, and you will see a screen like the one in the following screenshot, where you should click on **START TRIAL** first and then click again on **ENABLE**. If the mentioned buttons are no longer clickable or the screen does not appear, you may have enabled the API and the trial in the past, and you can safely skip this step:

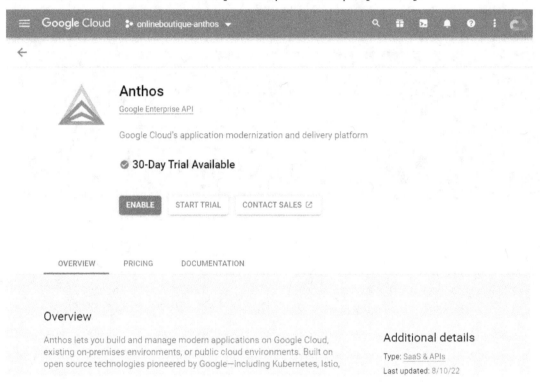

Figure 7.5 – Anthos trial and enable the API screen

Now click on the **Service Mesh** section on the left-side bar of Anthos and look for a table that includes all the different services for our online boutique on the right side of the screen: frontend, recommendations, checkout, currency, email, and payments.

Next, switch to the **Topology** view using the icon at the top-right side of the Anthos Service Mesh area, and we will be able to see how services are connected thanks to a graph view like this one:

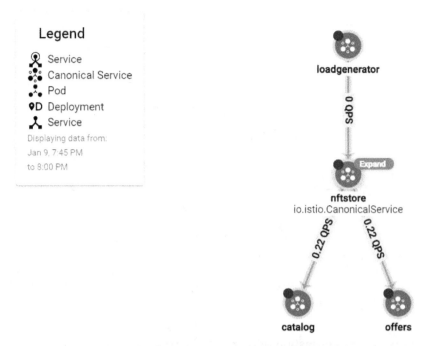

Figure 7.6 – An Anthos Service Mesh topology view

Notice how you can click on each node on the graph to see key metrics about each service on the right side of the screen. Also, if you move your cursor over a service, the graph will show the number of **queries per second** (**QPS**) the service sends to the rest it depends on. The preceding screenshot also shows these QPS numbers.

Apart from being able to generate topology maps, this view can also be used to display a current and historical snapshot of the topology, allowing us to see which changes were applied to the service over time and even compare snapshots of the service taken at separate times, as a kind of a visual service change log.

Now please go back to the **Table view** using the button on the top-right side and look for and click on the `catalog` entry in the list of services. Then, select the **Traffic** view using the corresponding option in the left-side menu. Notice how there is a single version of the `catalog` service, as shown in the following screenshot. On the right side, we can also see graphs for some of the key metrics of our service:

Figure 7.7 – Traffic view for productcatalogservice using a single version of the service

We will now deploy a version 2 of the `catalog` service and will send 25% of the traffic its way, with the other 75% still being routed to version 1. Please leave the **Traffic view** page open in a tab because we will come back to it right after the deployment.

This is what is called a **canary deployment**, where we only send part of the traffic to an updated version of a service so we can confirm that it works correctly before we approve it to fully replace the earlier version.

To do this, let's first create a **destination rule** for `productcatalogservice`:

```
kubectl apply -f yaml/destinationrule.yaml
```

Then let's create v2 of our `productcatalog` service:

```
kubectl apply -f yaml/catalog-v2.yaml
```

Finally, let's split the traffic with 75% for v1 and 25% for v2:

```
kubectl apply -f yaml/split-traffic.yaml
```

After deploying these new services, wait for three minutes, go back to the main page of the online store in your web browser, and reload it a few times. You will notice that it's slower now. The reason for this is that v2 of the `catalog` service added a latency of three seconds to each request. This means that if you reload any of the pages, you should receive a slower response one out of four times.

Now, please load a few more pages on the store to generate data. You should also notice the effect of the added latency during this process.

Then, just wait for a few minutes for the load generator component to make random requests and get the metrics populated. Finally, reload the tab where you had the **Traffic** view, and you will now see a traffic split in place, like the following:

Figure 7.8 – Traffic split view for the online boutique

Traffic is now being split between the two versions, sending approximately 25% of the requests to catalog-v2, and the other 75% to catalog, which is the name for v1.

Also, look at the latency graph in the right column and see how catalog has a latency of a bit less than two seconds, while catalog-v2has almost four seconds due to the extra lag added in the code:

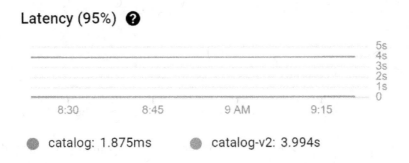

Figure 7.9 – The latency view for each version of the product catalog service

This is how we can identify an issue during a canary deployment and only affect a small percentage of our users. In cases like this, we would roll back the traffic split and remove catalog-v2 while we work on a fix.

This is the command to undo the traffic split, but there's no need to run it now:

```
kubectl apply -f yaml/rollback-traffic-split.yaml
```

And this is how we would delete `catalog-v2` if we found that it doesn't work as expected. Again, this is not needed right now, but you can look at the YAML file to have a better understanding of how deletions and rollbacks are implemented:

```
kubectl delete -f yaml/catalog-v2.yaml
```

Now, let's open Cloud Logging by visiting the following URL for your project:

```
https://console.cloud.google.com/logs/query
```

Since there is a load generator running in the background, there will be continuous requests to our online store, which we can take a look at. Your screen should look like this:

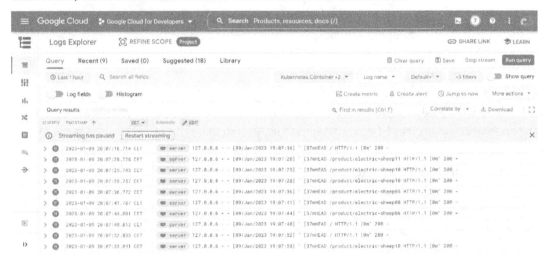

Figure 7.10 – Events for the online store in Cloud Logging

As you can see from this screen, it's easy to see what our applications running on Anthos are doing, even without debug logging enabled.

Now, try to filter the listing to display only error events:

1. Find the **Severity** drop-down list at the top-right part of the screen.
2. Check the box next to **Error** and click apply.

This is how to display only error messages. We can choose one or more severities, which is especially useful for troubleshooting. Notice how you can also filter by resource or log name using the other two drop-down lists next to the one for severity. Finally, notice how the first drop-down list on that same line allows us to select the time range that we want to use, including both default ranges and the possibility to define a custom one. Filters can be defined using the drop-down lists or by writing a query.

The following screenshot displays logs for our online store on a specific node. You can see the selections in the drop-down lists and how these are also shown as a query:

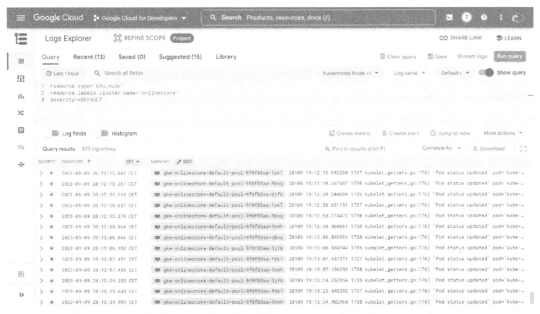

Figure 7.11 – Filtering events in Cloud Logging using drop-down lists or queries

Now let's check the metrics for our cluster by opening Cloud Monitoring using this URL: `https://console.cloud.google.com/monitoring`.

If you go to the **Metrics explorer** section using the left menu, you can click on the **Select a Metric** button and see that there are a lot of metrics to choose from. For example, choose **Server Response Latencies**, as shown in the following screenshot:

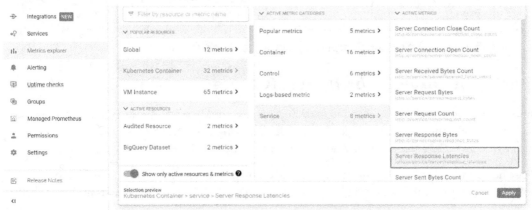

Figure 7.12 – Metric selection screen in Cloud Monitoring

Now choose to group by `destination_service_name`; which services have the highest latency? In my case, they are the `nftstore-external` and `catalog` services, as you can see here:

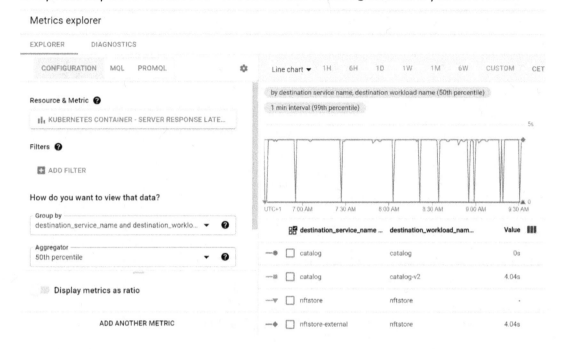

Figure 7.13 – A customized Metrics explorer graph for server response latencies

Take some time to display other metrics. As you can see, there are a lot to choose from, including metrics from Kubernetes, Istio, Networking, IAM, and metrics about consumed API calls and quotas, together with audit events. These can be a great aid in performance and cost optimization procedures.

I hope this exercise helped you better understand how Anthos works and how easily we can deploy services in Google Cloud. But Anthos also makes it easy to run our workloads on other providers. To prove it, let's use the next section to deploy our store on Microsoft Azure.

Running our example on Microsoft Azure

This section describes the steps required to deploy our store in Microsoft Azure, and it's based on the original example published by Google Cloud at the following URL: `https://cloud.google.com/anthos/clusters/docs/multi-cloud/azure/deploy-app`.

Please note how most of the commands are used to create the underlying infrastructure, but once it is available, the steps are the same as in any other platform, and all the setup is done from Cloud Shell.

To complete this part of the exercise, you need to have an active Microsoft Azure subscription. A free test account will not work due to the limitation in the number of cores that can be requested.

First, open Cloud Shell, set the PROJECT_ID shell variable, and check that it has the right value:

```
PROJECT_ID=<your-project-id>
echo $PROJECT_ID
```

Next, set which Cloud project to use with the following command. You may be asked to authorize the configuration change with your account:

```
gcloud config set project $PROJECT_ID
```

Then, use the following command to ensure that all required APIs are enabled:

```
gcloud --project="$PROJECT_ID" services enable \
  anthos.googleapis.com \
  cloudresourcemanager.googleapis.com \
  connectgateway.googleapis.com \
  gkemulticloud.googleapis.com \
  gkeconnect.googleapis.com \
  logging.googleapis.com \
  monitoring.googleapis.com
```

You should see a message saying Operation finished successfully.

Then, it's time to install the Azure **command-line interface** (**CLI**) utility by running the following command:

```
curl -sL https://aka.ms/InstallAzureCLIDeb | sudo bash
```

The installation will take a few seconds to complete.

You can verify that everything went well by issuing the following command, which will display a list with the version of each component used by the Azure client:

```
az version
```

Now, log on to your Azure account with the Azure CLI using the credentials that you obtained at the beginning of this section:

```
az login --use-device-code
```

Follow the instructions on the screen to authenticate, and you will see a JSON text, including information about your account. You can use the following command at any time to verify that your account was successfully added and show that JSON text again:

```
az account show
```

If you have more than one subscription and the one shown in the previous command is not the one that you want to use, just run the following command to list all available subscriptions:

```
az account list
```

And then copy the value for the `id` field of the right one, and enable it by issuing this command:

```
az account set -subscription <SUBSCRIPTION_ID>
```

You can then run `az account list` again to verify that the right subscription is now being shown.

We are now ready to start creating our infrastructure in Azure.

First, please use the following commands to choose which Azure and Google Cloud regions will host and manage your cluster. In my case, I chose `eastus` for Azure and `us-east4` for Google Cloud, but you may prefer to use other options that are closer to you. I recommend you choose regions that are geographically close to each other:

```
az account list-locations -o table
gcloud compute regions list
```

Then, we'll first define some environmental variables in Cloud Shell by pasting the following lines. These variables will be used in later commands.

Just replace the placeholders in the first and second lines with your chosen Azure and Google Cloud regions, respectively, and copy and paste the whole chunk into your console:

```
AZURE_REGION="<YOUR-CHOSEN-AZURE-REGION>"
GOOGLE_CLOUD_LOCATION="<YOUR-CHOSEN-GCLOUD-REGION>"
APPLICATION_NAME="azureapp"
CLIENT_NAME="anthosclient"
CLUSTER_RESOURCE_GROUP_NAME="azureapprg"
NAT_GATEWAY_NAME="azureappnatgw"
SSH_PRIVATE_KEY="./private_key.txt"
VNET_NAME="azureappvmnet"
VNET_ADDRESS_PREFIXES="10.0.0.0/16 172.16.0.0/12"
VNET_RESOURCE_GROUP_NAME="azureappvmnetrg"
```

Now, it's time to create the different resources in Azure.

First, let's create a new resource group for the virtual network:

```
az group create \
--location "$AZURE_REGION" \
--resource-group "$VNET_RESOURCE_GROUP_NAME"
```

Then, let's create a new virtual network with a default subnet. You can safely ignore the warning related to Microsoft.Network:

```
az network vnet create \
--name "$VNET_NAME" \
--location "$AZURE_REGION" \
--resource-group "$VNET_RESOURCE_GROUP_NAME" \
--address-prefixes $VNET_ADDRESS_PREFIXES \
--subnet-name default
```

Now, we will create an IP address for a new **network address translation (NAT)** gateway:

```
az network public-ip create \
--name "${NAT_GATEWAY_NAME}-ip" \
--location "$AZURE_REGION" \
--resource-group "$VNET_RESOURCE_GROUP_NAME" \
--allocation-method Static \
--sku Standard
```

Next, we will attach a NAT gateway to the IP address:

```
az network nat gateway create \
--name "$NAT_GATEWAY_NAME" \
--location "$AZURE_REGION" \
--resource-group "$VNET_RESOURCE_GROUP_NAME" \
--public-ip-addresses "${NAT_GATEWAY_NAME}-ip" \
--idle-timeout 10
```

Finally, we will attach the NAT gateway to the default subnet:

```
az network vnet subnet update \
--name default \
--vnet-name "$VNET_NAME" \
```

```
--resource-group "$VNET_RESOURCE_GROUP_NAME" \
--nat-gateway "$NAT_GATEWAY_NAME"
```

We are done with the networking part, let's work on other resources.

First, we will create a separate resource group for our Azure clusters:

```
az group create --name "$CLUSTER_RESOURCE_GROUP_NAME" \
--location "$AZURE_REGION"
```

Next, we will create an **Azure Active Directory (Azure AD)** application and principal, which Azure clusters will use to store configuration information. Let's first create the Azure AD application by issuing the following command:

```
az ad app create --display-name $APPLICATION_NAME
```

We will save the application ID in an environment variable for later use:

```
APPLICATION_ID=$(az ad app list --all \
--query "[?displayName=='$APPLICATION_NAME'].appId" \
--output tsv)
```

You can verify that the variable has a valid value using this command:

```
echo $APPLICATION_ID
```

Finally, we will create a service principal for the application:

```
az ad sp create --id "$APPLICATION_ID"
```

We will now create a few roles to allow Anthos clusters on Azure to access Azure APIs. First, let's define a few more environment variables that we will need to use in minute:

```
SERVICE_PRINCIPAL_ID=$(az ad sp list --all  --output tsv \
--query "[?appId=='$APPLICATION_ID'].id")

SUBSCRIPTION_ID=$(az account show --query "id" \
--output tsv)
```

Feel free to *echo* each of the variables to check for valid values.

In the next step, we will assign the `Contributor`, `User Access Administrator`, and `Key Vault Administrator` roles to our subscription:

```
az role assignment create \
--role "Contributor" \
--assignee "$SERVICE_PRINCIPAL_ID" \
--scope "/subscriptions/$SUBSCRIPTION_ID"


az role assignment create \
--role "User Access Administrator" \
--assignee "$SERVICE_PRINCIPAL_ID" \
--scope "/subscriptions/$SUBSCRIPTION_ID"


az role assignment create \
--role "Key Vault Administrator" \
--assignee "$SERVICE_PRINCIPAL_ID" \
--scope "/subscriptions/$SUBSCRIPTION_ID"
```

Anthos clusters on Azure use an **AzureClient** resource to authenticate.

To create one, let's first set the environment variable to hold information about our Azure tenant ID:

```
TENANT_ID=$(az account list \
--query "[?id=='${SUBSCRIPTION_ID}'].{tenantId:tenantId}" \
--output tsv)
```

Did you check that it was assigned a valid ID by using `echo` to display its value?

Then, let's create the actual client:

```
gcloud container azure clients create $CLIENT_NAME \
--location=$GOOGLE_CLOUD_LOCATION \
--tenant-id="$TENANT_ID" \
--application-id="$APPLICATION_ID"
```

Next, let's save the certificate to a variable, so we can upload it:

```
CERT=$(gcloud container azure clients get-public-cert \
--location=$GOOGLE_CLOUD_LOCATION $CLIENT_NAME)
```

And finally, let's upload the certificate to your application on Azure AD:

```
az ad app credential reset \
--id "$APPLICATION_ID" \
--cert "$CERT" --append
```

Now, we will create a public and private key pair to encrypt communications and associate it with our control plane and node pool virtual machines.

Run the following command to create the key pair. Notice that you may need to press the *Enter* key for the command to complete:

```
ssh-keygen -t rsa -m PEM -b 4096 -C "COMMENT" \
-f $SSH_PRIVATE_KEY -N "" 1>/dev/null
```

Let's next store the public key in an environment variable using the following command:

```
SSH_PUBLIC_KEY=$(cat $SSH_PRIVATE_KEY.pub)
```

Next, let's set the default management location for our Azure cluster. This is the reason I recommended you choose Azure and Google Cloud regions that are close:

```
gcloud config set container_azure/location \
$GOOGLE_CLOUD_LOCATION
```

Now, we will save our cluster's resource group to an environment variable by running the following command:

```
CLUSTER_RESOURCE_GROUP_ID=$(az group show \
--query id \
--output tsv \
--resource-group=$CLUSTER_RESOURCE_GROUP_NAME)
```

Next, we'll save our cluster's Virtual Network ID to an environment variable by running the following command:

```
VNET_ID=$(az network vnet show \
--query id --output tsv \
--resource-group=$VNET_RESOURCE_GROUP_NAME \
--name=$VNET_NAME)
```

Finally, let's save our cluster's subnet ID to an environment variable by running the following command:

```
SUBNET_ID=$(az network vnet subnet show \
--query id --output tsv \
--resource-group $VNET_RESOURCE_GROUP_NAME \
--vnet-name $VNET_NAME \
--name default)
```

All the preparations are now finished, and we are ready to create our cluster in Azure!

The following command will create our cluster in Azure. Notice which CIDR blocks I chose for pods and services and customize them to your requirements. I also chose our own project ID to register the cluster; please change it, too, if you prefer to use any other. Notice that the cluster version was the latest one available at the time of writing this chapter, and a new one will probably be available when you read it.

This command will take a while to complete, this is normal, and it can be a very good time to go for a coffee!

The command may display a `PrincipalNotFound` error due to replication errors, but that shouldn't be a problem; just be patient and let the command finish. If the command does not run properly on the first run, it may be that the certificate that we generated a few minutes ago is not ready for use yet or that some other replication has not yet finished. Just wait a minute and try again. You may need to delete the cluster before you can try to create it again. If that is the case, please refer to the *Cleaning up* section for more details:

```
gcloud container azure clusters create azure-cluster-0 \
--cluster-version 1.25.5-gke.1500 \
--azure-region $AZURE_REGION \
--fleet-project $PROJECT_ID \
--client $CLIENT_NAME \
--resource-group-id $CLUSTER_RESOURCE_GROUP_ID \
--vnet-id $VNET_ID \
--subnet-id $SUBNET_ID \
--pod-address-cidr-blocks "192.168.208.0/20" \
--service-address-cidr-blocks "192.168.224.0/20" \
--ssh-public-key "$SSH_PUBLIC_KEY" \
--tags "google:gkemulticloud:cluster=azure-cluster-0"
```

If the command finished successfully, you should see an output like this:

```
Created Azure Cluster [https://us-east4-gkemulticloud.
googleapis.com/v1/projects/anthos-azure/locations/us-east4/
azureClusters/azure-cluster-0].
NAME: azure-cluster-0
AZURE_REGION: eastus
CONTROL_PLANE_VERSION: 1.25.5-gke.1500
CONTROL_PLANE_IP: 10.0.0.4
VM_SIZE: Standard_DS2_v2
STATE: RUNNING
```

Next, let's create a node pool:

```
gcloud container azure node-pools create pool-0 \
--cluster azure-cluster-0 \
--node-version 1.25.5-gke.1500 \
--vm-size Standard_B2s \
--max-pods-per-node 110 \
--min-nodes 1 \
--max-nodes 5 \
--ssh-public-key "$SSH_PUBLIC_KEY" \
--subnet-id $SUBNET_ID \
--tags "google:gkemulticloud:cluster=azure-cluster-0"
```

Again, a successful execution will finish with a message similar to the following one:

```
Created Azure Node Pool [https://us-east4-gkemulticloud.
googleapis.com/v1/projects/anthos-azure/locations/us-east4/
azureClusters/azure-cluster-0/azureNodePools/pool-0].
NAME: pool-0
NODE_VERSION: 1.25.5-gke.1500
VM_SIZE: Standard_B2s
MIN_NODES: 1
MAX_NODES: 5
STATE: RUNNING
```

You can use the following command to check the status of your cluster:

```
gcloud container azure clusters describe azure-cluster-0 \
--location $GOOGLE_CLOUD_LOCATION
```

Once the cluster and the node pool are ready, we will just need to obtain application credentials for the cluster by running the following command:

```
gcloud container azure clusters \
get-credentials azure-cluster-0
```

The command will display an output like this one:

```
A new kubeconfig entry "gke_azure_anthos-azure_us-east4_azure-
cluster-0" has been generated and set as the current context.
```

Then, we will be ready to deploy our application using almost the same commands that we used for Google Cloud, which is the magic of Anthos. Please remember that these commands should be run from the base directory in the code repository for this chapter.

The only command that changes its format is asmcli, which must be run with the following command-line parameters from Cloud Shell, where you should replace <your_username> with your own username to provide a path to the configuration file. Also, make sure that asmcli was downloaded during the Google Cloud deployment exercise; otherwise, the execution will fail:

```
./asmcli install \
--fleet_id $PROJECT_ID \
--kubeconfig /home/<your_username>/.kube/config \
--output_dir ./asm_output \
--platform multicloud \
--enable_all \
--ca mesh_ca
```

Then, let's add a label using the commands that we already know:

```
REVISION=$(kubectl -n azure-cluster-0 \
get mutatingwebhookconfiguration \Ç-1 app=sidecar-injector \
-o jsonpath={.items[*].metadata.labels.'istio\.io\/
rev'}'{"\n"}' | awk '{ print $1 }')
```

And then the following:

```
kubectl label namespace default istio-injection- \
istio.io/rev=$REVISION --overwrite
```

Now, it's time to deploy our sample application:

```
kubectl apply -f yaml/kubernetes.yaml
```

Then, just wait for all the pods to be ready using the following command:

```
kubectl get pods
```

We should get two pods for each service due to the sidecar proxies, which will provide information about our services to Anthos Service Mesh.

The deployment is complete at this point. Now, you can find the external IP address of your store by running the following command, and you can visit the application frontend by pasting this IP address into your web browser:

```
kubectl get service nftstore-external | awk '{print $4}' | tail
-1
```

Visit `http://<YOUR-IP-ADDRESS>` and play around a bit with the website.

After a few minutes, our new cluster should appear in the Google Cloud UI section for Anthos, just as it did for its Google Cloud equivalent. This is what the cluster information section should look like:

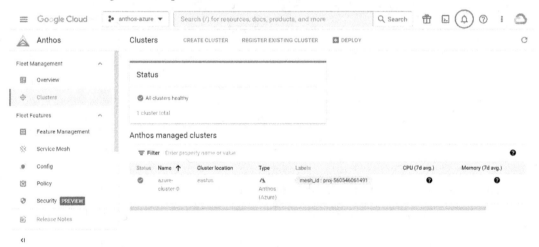

Figure 7.14 – Azure Kubernetes cluster information shown in the Anthos UI

And the Anthos Service Mesh will also provide us with visibility about the different services, even if this time they are running in Azure, as you can see in the following screenshot:

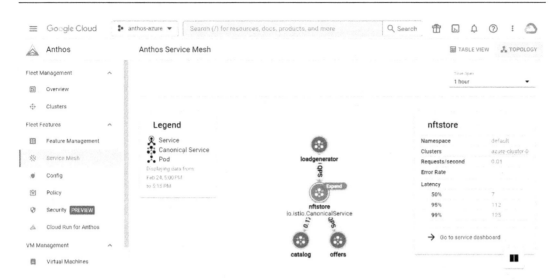

Figure 7.15 – The Topology view of our application from the Anthos UI

If you want to clean up once you have tested the application and played around with the Azure UI or its CLI, just use the following commands to delete all the created resources:

```
gcloud container azure node-pools \
delete pool-0 --cluster azure-cluster-0
gcloud container azure clusters delete azure-cluster-0
az network vnet delete \
--name "$VNET_NAME" \
--resource-group "$VNET_RESOURCE_GROUP_NAME"

az network nat gateway delete \
--name "$NAT_GATEWAY_NAME" \
--resource-group "$VNET_RESOURCE_GROUP_NAME"

az network public-ip delete \
--name "${NAT_GATEWAY_NAME}-ip" \
--resource-group "$VNET_RESOURCE_GROUP_NAME"
```

> **Note**
>
> This project uses a load generator, so leaving it running overnight will mean having continuous requests in our online boutique. Please take this into consideration since it may affect your available credits or incur costs. My recommendation is to clean up at once after completing the exercise.

Cleaning up

If you created a new project in Google Cloud to run this example, please delete it so you don't incur any extra costs. If you reused another project, you can just delete the cluster using the following command:

```
gcloud container clusters delete $CLUSTER_NAME \
--zone=$COMPUTE_ZONE
```

Now, let's summarize what we discussed in this chapter.

Summary

This chapter started by enumerating the key points to consider when choosing a cloud provider, together with the common pitfalls and second thoughts that we can have during and after the process, and how being able to work in different environments and providers simultaneously can be extremely beneficial.

Then we introduced Anthos, a platform that can help us use multiple environments and cloud providers while providing unified management, security, and observability capabilities.

After deep diving into Anthos components, concepts, and features, we had the chance to use a hands-on example to better understand the benefits of Anthos and deploy a test application to Google Cloud and Azure.

In the next two chapters, I will show you how Google Cloud services and APIs can help us simplify our code while bringing advanced and useful features to our applications and services.

Further reading

To learn more about the topics that were covered in this chapter, take a look at the following resources:

- *8 Criteria to ensure you select the right cloud service provider*: https://cloudindustryforum.org/8-criteria-to-ensure-you-select-the-right-cloud-service-provider/

- *A valuable resource containing information about how to choose an interconnect type is How to choose a Network Connectivity product*: https://cloud.google.com/network-connectivity/docs/how-to/choose-product

- *Fleet management examples*: https://cloud.google.com/anthos/fleet-management/docs/fleet-concepts/examples

- *NFTs explained*: https://www.theverge.com/22310188/nft-explainer-what-is-blockchain-crypto-art-faq

Part 3: Extending Your Code – Using Google Cloud Services and Public APIs

Another of the benefits of Google Cloud is that it provides a lot of products and services, which we can integrate with our existing applications and services to add new and powerful features just by adding a few lines of code.

In this part of the book, we will first cover how networking works in Google Cloud since it's a very important part of cloud architecture, and then we will describe different advanced services that we can use to make our code even better, including storage and databases, translation and text-to-speech services, and even artificial intelligence-based APIs, which allow us to easily analyze our media assets to understand their contents, opening the door to a lot of new and exciting applications.

This part contains the following chapters:

- *Chapter 8, Making the Best of Google Cloud Networking*
- *Chapter 9, Time-Saving Google Cloud Services*
- *Chapter 10, Extending Applications with Google Cloud Machine Learning APIs*

Making the Best of Google Cloud Networking

In the last four chapters, we presented different options for running our code, from the simplest to the most complex, and we ended our journey by discussing the hybrid and multi-cloud architectures facilitated by Anthos.

All these scenarios have something in common: we want to run our code on the cloud, take advantage of its many benefits, and expose it as a service or application to our users.

This chapter will present important concepts about networking that can help us improve the availability, performance, and security of our services and applications deployed on Google Cloud.

I will first enumerate and briefly describe all the network-related services available in Google Cloud. Then, we will discuss how resources are organized in Google Cloud data centers. In the following section, we will dive deep into a few services that can help us connect to our resources on Google Cloud.

Next, we will discuss some of the basic Google Cloud networking services, such as **Content Delivery Network (CDN)**, **Domain Name System (DNS)**, Load Balancing, and Cloud Armor. Finally, we will find out how the network service tiers can help us prioritize performance or cost savings, depending on our priorities.

We'll cover the following main topics in this chapter:

- Introducing Google Cloud networking
- Understanding regions and zones
- Connecting to our cloud resources
- Basic Google Cloud networking services
- Network Service Tiers
- Sample architecture

Introducing Google Cloud networking

Networking is a key component of any public cloud provider. Connecting our private resources to the internet in the best possible way, considering performance, latency, security, and availability will be key for our online strategy to succeed.

Google Cloud networking services have three key features that make them the best candidates for enterprises to use to modernize their infrastructure:

- They are built on top of a planet-scale infrastructure, making it possible to get closer than ever to our users, minimizing latency and maximizing performance
- We can leverage automation to make tasks easier and managed services that minimize administrative and maintenance tasks
- We can use AI and programmability to make services detect and react to specific situations, reconfiguring themselves automagically

With these three features in mind, network services and technologies in Google Cloud have been organized into four different families according to their purpose: connect, scale, secure, and optimize.

The *connect* family includes quite a few products. We will see some of them later, in their own sections.

The rest of the key services for connecting our resources to the outside world are as follows:

- **Private Service Connect**: This provides private connectivity to our own services, Google services, and third-party services from our **Virtual Private Cloud** (**VPC**).
- **Service Directory**: This is a managed service to publish, discover, and connect all application services by registering their endpoints. It offers inventory management at scale.

We will cover the *scale* family of products later in this chapter, including **Cloud Load Balancing**, **Cloud CDN**, and **Media CDN**.

Next is the family of products used to *secure* our network. While there will be a specific section dedicated to talking about Cloud Armor, there are a few other products worth mentioning:

- **Cloud IDS**: This is a detection service for intrusions, malware and spyware, and botnet attacks on our network
- **Cloud NAT**: This assigns internal IP addresses to our private resources while providing them with controlled internet access and preventing external access
- **VPC Service Controls**: These allow us to define security perimeters for API-based services, mitigating any potential opportunities for our data to be exfiltrated

And last, but not least, is the family of products used to *optimize* our network. **Network Service Tiers** will have its own section later in the chapter, but it's also worth mentioning **Network Intelligence Center**, a product that provides observability into the status of our network, including health, connectivity, performance, and security.

Having introduced the networking services available in Google Cloud, let's deep dive into how Google Cloud organizes its resources around the world.

Understanding regions and zones

When we use Google Cloud to deploy our code using any of the different options that we already discussed in earlier chapters, one of the most important decisions to make is where to deploy and run our workloads.

Compute Engine resources, such as virtual machines or containers, are available in data centers located in many different countries. Google Cloud organizes these resources geographically using regions and zones.

A **region** is just a geographical location that we can choose to host our cloud resources. Three examples would be *us-west-1*, *asia-northeast-2*, and *europe-west-3*. As you can see, the first part of a region includes the continent, followed by the area, and ends with a number that identifies the region.

Each region contains three or more **zones**, which are deployment areas for our cloud resources. Zones in each region are interconnected with low latency and high bandwidth.

For example, one of the previously mentioned regions, *asia-northeast-2*, contains the following zones: *asia-northeast2-a*, *asia-northeast2-b*, and *asia-northeast2-c*, all three located in Osaka, Japan. Zone names begin with the region name, followed by a letter identifying the zone.

If we have customers in or near Osaka, we should choose *asia-northeast-2* as the region to deploy our resources because then they will be geographically closer to our users. If we also want higher availability for our services, we should deploy these resources to different zones in that region, so that an outage in any of the zones will not affect our service.

We can achieve an even better availability level if we replicate our resources in multiple regions. Continuing with our example, if we want extra availability and we are planning to expand our business to Tokyo, or if we already have customers there, it would make sense to replicate our infrastructure to *asia-northeast1*, which also has three zones; all of them are located in Tokyo.

If our company keeps on growing and decides to open local headquarters in the EMEA region and the US, and we get a lot of new customers in those markets, we will probably decide at some point to deploy our resources to local regions in each of these continents so that we can reduce latency and meet legal requirements, again using different zones for better availability.

> **Note**
>
> Please take into account that while extending or replicating your architecture to multiple regions will make your architecture more robust, it may also require additional work to keep your data in sync, depending on the nature of your services, their requirements, and how they were implemented. This sync process will involve communication within or across regions, which will incur different costs. This added complexity and cost is another factor to take into account when deciding about both your location and expansion strategies.

Now that we know how to use regions and zones, let's talk about cloud resources and how they are bound to each other.

If we want to deploy a container, for example, it makes sense to think that it will run in a data center, and since there is at least one in each zone, we can say that a container is a *zonal resource*. If we attach a disk to this container, it will also be provided from a storage system within the data center so, again, a disk is also a zonal resource.

However, this container may have an external IP address associated with it. These addresses are shared between all the regions, which means that they are a *regional resource*.

As you can probably imagine, zonal resources can only be used within the scope of the zone where they are located, while regional resources can be requested and used in any of the underlying zones in that region.

For example, we can use an IP address that was just freed in the *asia-northeast1-b* zone in a container located in *asia-northeast1-a* because IP addresses are regional resources. However, if we try to attach a disk located in *asia-northeast1-c* to a virtual machine located in *asia-northeast-b*, that will fail because these resources are zonal, and we cannot attach them unless they are in the same zone.

Besides zonal and regional resources, we have also worked in previous chapters with **global resources**, which can be used in any region or zone. An example of a global resource would be a container image. Can you think of any other examples?

Finally, it's important to know that in Google Cloud zones and clusters are decoupled, which means that your organization may have its resources in one or more clusters within a chosen zone, while another may use a different set of clusters. This abstraction helps balance resources across the clusters available in a region and is also compatible with the addition of new regions as the available Google Cloud infrastructure grows over time.

Choosing the best region and zone

One of the steps that we need to complete before establishing our presence in the cloud is deciding where our resources will be geographically located.

As a rule of thumb, we should choose a region that is geographically closer to our customers. There are several reasons for doing this:

- **Lower latency**: Being physically closer to our customers reduces the latency to our workloads, which equates to better performance for our applications or services.

- **Higher availability**: If our services are replicated in different data centers, it will be easier to tolerate outages, either by redirecting traffic to others that are not affected or by having backup services ready to be run on different data centers.

- **Better compliance**: Some services need to follow strict regulations and policies, often preventing personal data from leaving the country where it was obtained. Being present in different countries can make it easier for us to comply with all these requirements.

However, there are some other important criteria to consider when making this choice:

- **Service and feature availability**: Not all regions have the same services available to Google Cloud customers. For instance, if we need to use GPUs for complex calculations or to train ML models in a shorter time, we need to choose a region where they are available. This is also applicable to virtual machine families, and we should not only consider current needs but also future ones before we make a final choice.

- **Carbon footprint**: Google is making a huge effort to minimize its footprint but, at the time of writing, some regions still have a higher carbon footprint than others, and you may want to exclude those from your list of candidates.

- **Pricing**: Services don't have the same cost in all regions, and if your budget is limited, you may prefer to sacrifice a bit of latency for a significant price cut. Please take into account that this will depend a lot on which services you use and how often you use them, and may not be worth it in many cases.

- **Quotas**: Google Cloud has both project-wide and per-region quota limits for certain resources. Being aware of these limits is essential if you want to make a good choice considering current usage and future plans for growth. You can find details about these limits on the following documentation page: `https://console.cloud.google.com/iam-admin/quotas`.

- **Expansion plans**: If you have a clear plan to extend your service to other countries or even continents, it may be worth building an architecture that supports multiple zones and regions from day one. Not taking this into account early enough is a common mistake that costs organizations a lot of money and time later. As we discussed earlier in the book, thinking big won't hurt anyone, and designing an application with scalability in mind from day one can only save you time and money in the future.

- **Service-Level Agreements (SLAs)**: If your service includes some kind of SLA, it is a good idea to have backup infrastructure ready in passive mode in another region or zone, ready to be awakened if an outage happens. This is just another kind of replication, but one that can save you a lot of money if a disaster happens.

You can find a list of zones and information about their available machine types, CPUs, carbon footprint levels, and whether they support GPUs on the following documentation page: `https://cloud.google.com/compute/docs/regions-zones`.

Next, let's discuss which options are available to connect Google Cloud with the rest of our infrastructure.

Connecting to our cloud resources

Before learning more about the connectivity options available in Google Cloud, let's first understand how networking works internally by introducing VPC networks.

VPC networks

Just as a Virtual Machine is the virtual version of a physical host, a VPC network is the virtual version of a traditional physical network. VPCs are directly implemented inside Google's production network.

A VPC network provides different services and features:

- Connectivity for our VMs and GKE clusters, App Engine flexible environment instances, and other Google Cloud products built on VMs.

- Native internal TCP/UDP load balancing and proxy systems for internal HTTP(S) load balancing.

- Compatibility with Cloud VPN and Cloud Interconnect to help us connect to our on-premises networks. We will cover these two products in a minute.

- Traffic distribution from Google Cloud external load balancers to backends.

Beyond this feature list, VPC networks support standard networking protocols and services such as IPv4 and IPv6, subnet creation, route definitions, and firewall rules, among others. We can even use a **shared VPC** to let multiple projects communicate with each other with higher standards of security and better performance.

A Google Cloud project can contain multiple VPC networks and, by default, each new project comes bundled with a VPC network that has one subnet. Regarding its scope, VPC networks are global resources while subnets are regional.

We can even connect one of our VPC networks to another located in a different Google Cloud project or organization by using **VPC network peering**.

In summary, VPC networks will cover all our internal networking needs, but let's now go through the different Google Cloud networking options to connect our cloud resources with on-premises networks, third-party data centers, or public cloud providers.

Network connectivity products

Google Cloud offers various products depending on our purpose and requirements.

Starting from the simpler to the more complex cases, if we just need to connect to Google APIs, also including Google Workspace if our organization uses it, these are our two options:

- **Direct Peering**: This establishes a bi-directional direct path from our on-premises network to Google's edge network, providing connectivity with any Google Cloud products and services that can be exposed through one or more public IP addresses. This option is available at more than 100 locations in 33 countries.

- **Carrier Peering**: This uses a service provider to set up a dedicated link that connects on-premises systems with Google, with a lower latency and higher availability.

If we want to access our VPC networks from either our on-premises networks or other cloud providers, we have three different options:

- **Cloud VPN**: This uses an IPSec VPN connection in a single region to securely connect our on-premises network to our VPC networks in Google Cloud, with an SLA of 99.99% service availability. This is the basic option.

- **Dedicated Interconnect**: The first Cloud Interconnect option creates a dedicated connection between our on-premises network and our VPC networks in Google Cloud. This is the most powerful option, a cost-effective choice for high-bandwidth needs, offering 10-Gbps or 100-Gbps circuits with attachment capacities from 50 Mbps to 50 Gbps. If you don't need that circuit speed, you can opt for Partner Interconnect.

- **Partner Interconnect**: The second option with Cloud Interconnect uses a service provider to set up a connection between our on-premises network and our VPC networks in Google Cloud. It offers flexible capacities from 50 Mbps to 50 Gbps and can be considered an intermediate option.

Please take into account that **Cloud Interconnect** is recommended as a better alternative to any of the peering options previously described as long as you don't need access to Google Workspace.

We can also use Google Cloud to connect our sites using **Network Connectivity Center**, which uses Google's network as a **Wide Area Network** (**WAN**) and reduces operational complexity using a hub-and-spoke model. Network Connectivity Center acts as the hub, and we can set up spokes so that different types of Google Cloud resources can be attached. For example, we can set up a spoke with a VLAN attached to one of our sites.

Finally, if we need to connect to external **CDN** providers, **CDN Interconnect** can help us optimize our CDN population costs while providing direct connectivity to select CDN providers from Google Cloud.

Now that we have been familiarized with the available options for connectivity, let's use the next section to discuss some of the basic networking services available in Google Cloud.

Basic Google Cloud networking services

Google Cloud offers other managed networking services with advanced features and little to no maintenance overhead. Let's cover the most important ones.

Cloud DNS

The internet was built on top of **Transmission Control Protocol and Internet Protocol (TCP/IP)**. IP makes it possible to assign each host a unique address, also known as an IP address, while TCP takes care of transmitting the information between two IP addresses. We can think of IP as a gigantic phone book and TCP as a big team of couriers responsible for delivering information from a source host to a destination.

There are two versions of the IP protocol that co-exist at the time of writing. The older one, IPv4, provides a 32-bit pool with up to 4,294,967,296 different addresses, most of which are currently in use. A newer version, IPv6, was built to solve the address shortage and added some interesting new features, providing a 128-bits pool with up to 340,282,366,920,938,463,374,607,431,768,211,456 addresses.

As you can imagine from these numbers, a *phonebook* is vital to help us connect to the right destination without having to remember thousands of numbers, and that's the reason domain names were created.

For example, if I want to take a look at the NBA's website to read the latest news on my favorite US basketball teams, I should remember their IP address, 34.213.106.51, so I can load their web page in my web browser. This would become rather difficult as I started to access many other websites. However, remembering its associated domain name, nba.com, seems much easier. Besides, the IP address associated with this domain name may be subject to changes; for example, if they move to a new provider or if good old IPv4 is deprecated at some time soon. The domain name, however, will remain the same for much longer.

As the NBA did some time ago, if we want to set up our presence on the internet in general, and if we need to use cloud resources in particular, we will need to use domain names to make it easier for our users and customers to connect to our online services. And here is where **Cloud DNS** comes to the rescue.

Cloud DNS is Google's implementation of a DNS, that is, a managed service used to publish information about our domain names to the global DNS, a hierarchical distributed database holding information about all the domains available worldwide, including IP addresses and other types of information provided by domain owners. Yes, this would be that gigantic phonebook we recently mentioned.

As owners of one or more domains, we will maintain a **DNS zone**, which is a small part of the global database where we will keep our names and IPs up to date. A managed zone contains DNS records with a common name suffix, such as nba.com. The combination of millions of DNS zones makes up the global DNS database, which is replicated multiple times by service providers using caching DNS servers, which provide a faster local experience to customers all around the globe.

All domain names are registered in a **domain registrar**, which provides information about the owner, together with a list of the IP addresses of the DNS servers, which can answer requests about the DNS zone for a specific domain. **Google Domains** (`https://domains.google`) is an example of a registrar, among many other ones available, while Cloud DNS is one of the many options available to host the information about our DNS zone.

As with other managed services, Google provides a UI and an API, which makes it much easier for us to use the service while they take care of DNS server administration and patching. Users can then perform lookup requests to get the IP addresses of our hosted services so that they can connect to and use them.

Cloud DNS supports both public and private DNS zones. The former are those that are visible to anyone once they have been published. The latter, however, will be internal and only visible from one or more VPC networks that we configure to have access. These are mostly used to provide **DNS resolution** for internal corporate IP addresses, which don't have any use outside our organization.

An interesting benefit of using a cloud-based DNS service is that our zonal databases are replicated across Google's regions, and any received requests are routed and responded to from the location closest to the requestor using **anycast**, which means that our users will experience lower latency and better performance.

Another nice feature of Cloud DNS is that it integrates with **Identity and Access Management** (**IAM**), allowing us to define access permissions at both the project and zone level, and is also compatible with the use of service accounts. This is key to ensuring that we can control who can make changes to our DNS zones.

A DNS system provides different types of records, which can be queried to obtain information that helps clients connect to servers. Each record has a type, which defines what it can be used for, an expiration date to indicate for how long the provided data is valid, and the actual data associated with the record type.

These are the record types that can be stored in a DNS zone for a domain in Cloud DNS:

- **A**: An address record that maps host names to their IPv4 address – for example, `65.1.4.33`, a bunch of numbers that we could potentially remember.

- **AAAA**: An IPv6 address record that maps host names to their IPv6 address. An example would be `c109:56a9:bfa0:9596:b84e:be91:875f:5dd6`. As you can see, these are virtually impossible to remember.

- **Canonical Name** (**CNAME**): A CNAME record that specifies alias names. For example, `nba.com` could be set up as an alias of `www.nba.com` so that all users writing the short name in their browsers are redirected to the full WWW domain.

- **Mail exchange (MX)**: An MX record that is used in routing requests to mail servers. It can be used to obtain the IP addresses of servers that handle incoming emails. One of the records for `nba.com` is `mxa-001bf601.gslb.pphosted.com. (148.163.156.86)`.

- **Name server (NS)**: A NS record that delegates a DNS zone to an authoritative server. It contains two or more entries for servers that can provide authoritative answers to requests about this zone. One of the records for `nba.com` is `a7-67.akam.net. (23.61.199.67)`.

- **Pointer (PTR) record**: A PTR record defines a name associated with an IP address. Is often known as a **reverse lookup**, which provides a name given an IP address. For example, if we looked up an IP address of `34.213.106.51` and it had a PTR record defined (which it doesn't), it would point to `nba.com` because this is another valid IP for NBA's website.

- **Start of authority (SOA)**: An SOA record is used to designate the primary name server and administrator responsible for a zone. Each zone hosted on a DNS server must have an SOA record and this is useful for obtaining more information about who owns a domain. In our example, the SOA record for `nba.com` defines `a1-148.akam.net` as the primary NS and `dnsteam@nba.com` as the responsible email. An SOA record also has a serial number that increases with any change performed in the record, so it's easier to ensure that DNS caches always have the latest version.

As you can see from this list, we can use DNS records to resolve the IP address of a specific domain name (both for IPv4 and IPv6), get the IP addresses where our emails should be sent for a specific domain, or get a list of the primary name servers for a specific zone. All these requests, properly combined, can provide us with all the information we need at the IP protocol level to connect to a specific address. Then, the TCP protocol will be used to route the packets containing our information to the IP address indicated.

DNS information is critical in terms of security. If someone can point our domain names to another IP address, they can set up a fake server and impersonate ours, which could cause us serious problems. To prevent this, the **Domain Name System Security Extensions (DNSSEC)** can be used to authenticate responses provided to lookups by the DNS system. These extensions are only available in public DNS zones.

For the DNSSEC validation process to work properly, all elements in the lookup chain must be properly configured: the DNS server must be compatible with it, the specific zone must have a DNSKEY record, and the client must enable authentication in the lookup request.

Security and observability in Cloud DNS are provided via integrated monitoring and logging, which includes audit information, together with a dashboard where we can see graphs and metrics about latency, queries per second, or error rates, which can help us detect and troubleshoot any issues.

In the next section, we will discuss how load balancing is implemented in Google Cloud, a key service if we want to deploy services for many users.

Cloud Load Balancing

A load balancer distributes user traffic across multiple instances of our applications or services, thus reducing the risk of performance issues caused by sudden traffic peaks.

Cloud Load Balancing is a distributed managed service providing virtual load balancers that minimize management tasks, offering low latency and superior performance and supporting more than 1 million queries per second.

There are different load balancer types we can choose from, so there are a few questions that can help us choose the best one for our needs:

- What type of traffic do we need to balance? There are Layer 4 load balancers available for HTTP, HTTPS, TCP, SSL, UDP, ESP, and ICMP.

- Do we need global or regional balancers? Global balancers distribute traffic to backends across multiple regions, while regional ones work with backends located in a single region. Please notice that global load balancing will only be available if we use the Premium Network Service Tier. We will discuss tiers later in this chapter.

- Is the expected traffic external or internal? External means that traffic is coming from the internet, while internal means that traffic begins and ends inside Google Cloud.

- Do we prefer a proxy or a pass-through balancer? Proxy balancers terminate client connections and open new ones to the target backends, while pass-through ones preserve client packet information and implement direct server returns, where the response doesn't go back through the load balancer, but directly to the client.

Google Cloud offers software-defined load balancers with no hardware involved, which reduces maintenance needs and uses a single anycast IP address to offer a global frontend available in all regions worldwide. Seamless autoscaling makes it possible to handle unexpected traffic peaks without pre-warning. Also, Layer 7 load balancers can use advanced routing, such as using the URI or a field in the HTTP header to decide which is the best backend to use.

Cloud Load Balancing has native integration with Cloud CDN for delivering cached content and Cloud Armor for advanced protection.

Speaking of the devil, next, we will discuss how Cloud Armor can help us prevent attacks and mitigate online threats.

Cloud Armor

When online services are exposed to public connections and traffic, either directly or using load balancers, these services should be properly protected from potential threats that may affect their performance or compromise their security.

These are some of the most common dangerous attacks:

- **Distributed denial-of-service (DDOS)** attacks aim to disrupt the normal behavior of a service by flooding it with internet traffic coming from thousands of different hosts, making the target service unavailable

- **Cross-site scripting (XSS)** attacks are used to send malicious code, generally using browser-side scripts, to end users of a web application

- **SQL injection (SQLi)** attacks use bad filtering of parameters in the code to inject additional SQL clauses, allowing the attacker to gain access to privileged data or even delete the contents of a database

- **Local file inclusion (LFI)** attacks trick a web application to display local files, either exposing sensitive information or remotely executing unwanted code

- **Remote file inclusion (RFI)** attacks exploit web applications that dynamically include or reference external scripts or files, tricking them into loading different files instead, which often contain malware

- **Remote code execution (RCE)** allows attackers to run malicious code remotely by using exploits

Any public service is exposed to potential risks and threats and should be actively protected and monitored from a security perspective. **Google Cloud Armor** is a security product built for this purpose and is able to detect and mitigate the aforementioned risks and also other attack types working at the load balancer level. It's compatible not only with native applications running on Google Cloud but also with hybrid and multi-cloud architectures.

Cloud Armor is compatible with the following types of load balancers:

- A global external HTTP(S) load balancer
- A global external HTTP(S) load balancer (classic)
- An external TCP proxy load balancer
- An external SSL proxy load balancer

Security is provided using a sophisticated **Web Application Firewall (WAF)**, which includes thousands of complex preconfigured policies, each with dozens of rules compiled from open source security standards. Some of these rules only apply to HTTP(S) load balancers since some of the most popular threats previously mentioned only target web applications.

We can also add our custom policies using a specific language that can help us define which match conditions will trigger the policy and which actions will be taken when in that case.

Additional security capabilities include **rate limiting**, which can be used to prevent the exhaustion of application resources; compatibility with **named IP lists**, provided and maintained by third-party providers and which can be associated with specific actions; and **Adaptative Protection**, which analyzes incoming traffic, detecting potential threats and dynamically applying WAF rules to mitigate them.

Cloud Armor is offered in two different service tiers:

- **Google Cloud Armor Standard**, which uses a pay-as-you-go pricing model and includes always-on DDoS protection and preconfigured WAF rules for **Open Web Application Security Project (OWASP)** Top 10 protection.

- **Managed Protection Plus**, a monthly subscription with all the features of Google Cloud Armor Standard, adding third-party named IP address lists, and Adaptive Protection. Subscribers also get access to DDoS bill protection and DDoS response team services.

Now, it's time to talk about the last networking service, in this case, used for content delivery.

Cloud CDN

Globalization is unstoppable. A few decades ago, it was inconceivable that a small store in Madrid, Spain, could be selling its products to users in Africa, South America, or Japan, at the same time. The internet has become a window to the world; companies no longer think locally and must be prepared to do business on a global scale.

Continuing with the example, if a well-known t-shirt company based in Madrid wants to sell its products worldwide, it will probably start by setting up an e-commerce website. There, it will organize its t-shirts by size, style, and color... and it will add high-density pictures (in terms of pixels) for each article so that customers can see all their details and fall instantly in love with them. It will need some JavaScript to implement fancy zoom effects on the pictures and a bunch of CSS style sheets to make the website look great, and it will be ready to rock!

If its web server is located in a hosting company, it will load blazingly quickly when a laptop is used to test it out, and the t-shirt company will be very happy with the result, but it won't be long before it gets an email from an angry customer in Japan, complaining because the t-shirts are great, but the website is really slow and browsing the product section is disappointing. How can this be possible? What can be done to fix it?

The bigger the distance between the user and the server, the bigger the latency is too, and the lower the speed, unless our hosting company has good peering and optimized routes for international traffic. In a global world, it's impossible to find a location that will work well for all our international customers. And this is where CDNs come to the rescue.

Our cloud resources should be as close as possible to our users, but what happens when our users can be anywhere in the world? The answer is quite simple: our resources should also be all around the world so that we can provide the best possible experience to potential customers wherever they are.

This approach was impossible due to costs not so long ago: setting up servers in many distinct locations would be extremely expensive and would take a lot of maintenance work to keep systems patched and files synchronized across locations.

Thanks to CDNs, we can now straightforwardly do this. Cloud CDN uses Google's global **edge network** to cache and serve content to users from the closest location, improving their experience by minimizing latency while maximizing bandwidth.

Caching means that CDNs intercept an object when it is requested, for example, an image in a website, retrieving it from the origin and storing a local copy before returning it to the requestor. Future requests will be directly served from the cache, saving time and improving speed because the content will be available in a location closer to the user. A cache is a storage area where we store copies of all these requests, together with information about their original location, so we can reuse them in future requests.

In our example, imagine that we have a CDN enabled in Japan. A Japanese user finds our store by searching on Google and gets to the product page of the t-shirt website, which has a lot of pictures of assorted designs. These images will be configured to load from Cloud CDN, which will detect that the user is based in Tokyo and will look for the closest location of a Google Cloud CDN, which is Tokyo in this case.

When each image is requested, Cloud CDN will check whether it already exists in the local cache. If it does, which is called a **cache hit**, the image will be returned directly from the cache.

If the image didn't exist in the local cache, which is known as a **cache miss**, the image will be loaded and copied into the local cache from the origin server, which can be either the original web server or a Google Cloud Storage bucket, depending on our configuration, and will be returned to the user from the local cache and stored for future requests.

The **cache hit ratio** is the percentage of requests that resulted in a cache hit, which means that the requested objects were served from the cache. This is a useful metric to understand how well our caching strategy is working.

As traffic grows from a specific location to our website, all users can benefit from cached files and have a better user experience while loading objects originally located in a remote location. The price for bandwidth is much cheaper too since hosting plans often have limited bandwidth and as the number of daily visitors increases, costs will rocket. The web server load is also much lower once we hand over the serving of popular and cacheable content to a CDN server. Indeed, we can cache all kinds of static assets such as images, videos, JavaScript, and CSS files, among many others.

Caches are built to be always full, so cache content needs to be constantly evicted. **Eviction** is the mechanism used to make room for updated content by deleting unpopular or old files. Since these are the first on the list to be evicted, they can be replaced with content that is newer and has a higher probability of being reused; objects may also be chosen at random.

If we are using content that needs to be periodically refreshed, imagine, for example, a section with a picture called "*Today's offer*" that changes every day, we can use headers in the origin server to define an **expiration date** for an object. The object's expiration date will be checked on the next request and if it has already expired, a new copy will be cached and the outdated one will be evicted.

Objects will be inserted in the cache when they are fetched from the origin. In some cases, you may need to force the eviction of some of these objects to prevent serving stale content, for example, out-of-date JavaScript or CSS. This can be done by following a process known as **cache invalidation**, which uses a path pattern to find which matching objects should be evicted.

Finally, an interesting practice is **negative caching**, where we configure our CDN to cache error responses too, preventing malformed requests or URLs no longer working, but still referenced in our website, from sending huge loads of traffic to the origin server, which would defeat the main purpose of a CDN.

Negative caching mitigates the effect of errors on the web server load at the cost of having to move error analysis from the original server to the cloud, which in our case is not a big issue thanks to **Cloud Logging**.

CDNs are easy to learn about but not so easy to master, but once they are properly tweaked, they can be a significant help to reduce load and minimize the cost of serving static assets while providing a great user experience wherever our users are located.

The same week that I was writing this chapter, Google Cloud announced the availability of a new product called **Media CDN**, specifically created for media streaming. This product complements Cloud CDN by providing optimized caching support for streaming video, large file downloads, and other similar workloads that require high-throughput egress. Media CDN can serve cached content using encrypted traffic (HTTPS) and supports **bring-your-own** (**BYO**) domains to replace Google-hosted default ones. It can also use Cloud Armor to control access to content, and also supports using a Private Cloud Storage bucket as the origin by only allowing authenticated requests to access its contents. You can read more about Media CDN here: `https://cloud.google.com/media-cdn`.

And this was the last service in this section. Finally, let's discuss how network tiering works in Google Cloud.

Network Service Tiers

Each organization has different use cases and requirements for its presence in Google Cloud. We will use different services, deployed in various locations, and will choose our own payment options.

Why should we not be able to do the same when it comes to networking? Indeed, Google Cloud was the first major cloud provider to offer a tiered model for networking, allowing customers to optimize for either performance or cost, depending on their needs.

To make things simple, the two available tiers are called **Standard Tier** and **Premium Tier**.

The main difference between them is that in the Standard Tier, the traffic between the internet and VM instances in our VPC network is routed over the internet in general, while in the Premium Tier, traffic is kept within Google's network as much as possible, which means better performance, obviously at a higher cost.

The Premium Tier routes traffic using Google's network until it reaches the user's ISP while for the Standard one, traffic will exit much earlier and will transit multiple networks, with higher latency and lower performance.

For this reason, the Standard Tier is recommended for services hosted within a region and provides a performance level like any other provider, while the Premium Tier is a much better choice for globally available services, with higher performance when compared with other providers.

As I already mentioned in other sections, the two network service tiers available are not mutually exclusive, and while it makes sense to use the Premium Tier for global and mission-critical systems, the Standard one works well for regional services without incurring unnecessary additional costs.

Standard Tier is also not compatible with services such as Cloud CDN or Cloud VPN/NAT gateways, as I already mentioned, with HTTP(S) balancers being only usable if the backend is a Cloud Storage bucket.

Whatever tier we use, all the traffic between our Virtual Machines will be kept on Google's network, regardless of whether they are in the same or different regions, whether a load balancer is on the path, or whether we are using public or private IP addresses.

> **Note**
>
> Standard Tier is only available to resources that use regional external IP addresses. Besides, it is only available in some of the Google Cloud regions, so you'd better check its availability if you are planning to use this tier.

That's everything on Network Service Tiers in Google Cloud. Next, let's look at a sample network architecture including many of the services we just discussed.

Sample architecture

After quite a few pages describing the different networking services and products, I thought that a wonderful way to finish the chapter would be to combine a few of them in a real word scenario.

Imagine a company called **e-Design 4U**, based in Portland, Oregon. It sells graphic designs and templates for web designers in the United States, but most of its customers are in the east and the west of the country.

The company wants to provide fast downloads from both sides of the country but would like some added capacity to alleviate the load on its private servers a bit, located in a hosting provider in Portland.

Security is also important, and the company wants to make sure that it's not possible to access its content without a paid subscription.

Given this scenario, we could think of an architecture combining some of the services we covered in the chapter. Let's take a look at the following diagram and then we will discuss how it works:

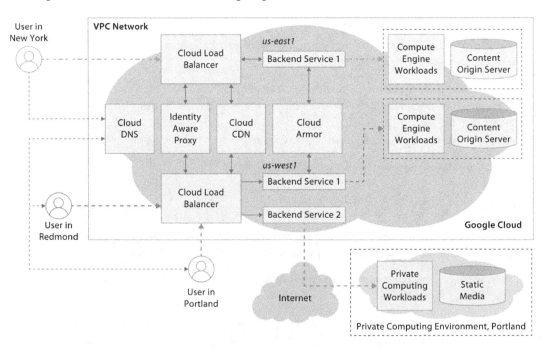

Figure 8.1 – Architecture diagram for our sample networking architecture

The proposed design uses similar workloads on both sides of the United States, in the *us-east1* and *us-west1* regions, to create a hybrid architecture. A VPC is used to connect both regions, making it easier to replicate media and to use a common set of tools and processes across the whole Google Cloud environment, regardless of the region.

Cloud DNS will translate the domain name to the IP address of the load balancer when users want to access the website by typing www.edesign4u.com in their browsers, the domain name used by the company. Cloud Load Balancing will then take users to either the east or the west cloud region, depending on their location. Users who are geographically close to the hosting provider, such as the example one in Portland in the earlier figure, will be connected directly to the private hosting environment, but the rest will be balanced across regions, where an elastic setup will be able to scale the number of nodes up and down, depending on the traffic received.

Static media will be originally hosted in Portland but will also be replicated in the Google Cloud regions using a local media storage component to store a copy of the files and Cloud CDN to cache recently requested files closer to the edge to minimize latency and maximize download speeds.

Notice how load balancers are connected to an **Identity-Aware Proxy** (**IAP**) to ensure that only authenticated users can access the private part of the website. The company only creates accounts for users with active subscriptions, using this system to protect the media files that they produce. The private part of the website also allows subscription renewals.

Finally, Cloud Armor is enabled to include firewall rules to allow traffic from the IP addresses of the load balancers, denying any traffic. CDN traffic will not pass through Cloud Armor, but any requests that reach the workloads or duplicate media content servers will be analyzed to minimize intrusion and any other potential security risks.

I hope you found this example interesting. Now, as an optional exercise, please take a few minutes to think how you would extend this scenario to other parts of the world, maybe using Tokyo or Amsterdam as the new headquarters for Asia and Europe, respectively. Also, how would you change the design if the company wants to stop using its hosting provider and run its website on Google Cloud?

Once you have answered those questions, it's time to wrap up.

Summary

This chapter started with a brief introduction to networking in Google Cloud. Then, we discussed how regions and zones work and then explained how we can connect to our cloud resources.

Right after, we deep-dived into some of the most important networking services available in Google Cloud, including Cloud DNS, Load Balancing, Cloud Armor, and Cloud CDN.

Finally, we talked about how the two different Network Service Tiers can be used to prioritize performance or cost savings and analyzed a sample architecture showcasing many of the network services and products discussed during this chapter.

In the next two chapters, I will show you how Google Cloud services and APIs can help you simplify your code while bringing advanced and useful features to your applications and services. Then, it will be the perfect time to put everything we have learned into practice.

Further reading

To learn more about the topics that were covered in this chapter, take a look at the following resources:

- *DNS Security Extensions (DNSSEC) overview*: `https://cloud.google.com/dns/docs/dnssec`

- *Best practices for Cloud DNS*: `https://cloud.google.com/dns/docs/best-practices`

- *Load balancer feature comparison*: `https://cloud.google.com/load-balancing/docs/features`

- OWASP list of attacks: `https://owasp.org/www-community/attacks/`

- *Google Cloud Armor Adaptative Protection overview*: `https://cloud.google.com/armor/docs/adaptive-protection-overview`

- *Network Service Tiers overview*: `https://cloud.google.com/network-tiers/docs/overview`

9

Time-Saving Google Cloud Services

The time available to complete a migration or develop a new application is often too short. Using some of the services provided by Google Cloud as building blocks for our applications can save us time and money when creating new services or migrating legacy ones. Even if we are not short of time, using managed services will help us mitigate security risks and reduce maintenance tasks.

Let's introduce some of the most useful Google Cloud services for developers and use an example to understand how we can use them together.

We'll cover the following main topics in this chapter:

- Cloud Storage for object storage and retrieval
- Cloud Tasks for asynchronous task execution
- Firestore in Datastore mode for high-performance NoSQL storage
- Cloud Workflows for service orchestration
- Pub/Sub for inter-service messaging
- Secret Manager for storing sensitive data
- Cloud Scheduler for running jobs at the right time
- A hands-on exercise to put them everything into practice

Let's get started!

Cloud Storage for object storage and retrieval

Our applications or services often need to store data files, and this can be a common source of headaches when our users are in different parts of the world. Luckily for us, Google Cloud has an ace in the hole to make our lives easier.

Introducing Cloud Storage

Cloud Storage is a managed service that allows us to store any amount of unstructured data and retrieve it as often as we like. Combine this with dual-region storage, turbo replication, automatic class assignment, and life cycle management, and this service can help us integrate advanced storage capabilities into our application while enormously simplifying our development cycle.

Cloud Storage stores immutable pieces of data, called **objects**, each of which can contain a file of any kind. We can organize our files in directories, and all of these are stored in **buckets**, which are containers associated with a Google Cloud project. Note that directories are just logical organization systems that make Cloud Storage more similar to how operating systems group files, but in reality, all objects in a bucket are stored at the same level. This is something to consider when we write code because a simple operation such as moving a lot of files from one directory to another is no longer as simple as updating the pointer. In this case, all objects will need to have their location metadata updated, which is a much more complex process. Buckets are resources that can be created in one or more regions and are integrated with **IAM**, allowing us to configure who can read, create, update, or delete them, and even making it possible to mark a container as public so that anyone can read its contents. Indeed, one of the uses for Cloud Storage is to implement a simple web server for static assets or to store the files for services such as Cloud CDN.

When we have access to a bucket, we can upload or download files to it, depending on our permissions, and there are different ways to do it. Cloud Storage is easy to use from the Cloud console for simple operations, and the **gcloud** utility can also be used to upload, copy, move, or delete big amounts of files using the command line. We can also integrate this service with our code using either its **REST API** or different **client libraries** provided by Google for different programming languages. We will see some examples in our hands-on exercise.

And if you remember, when we spoke about **Cloud Functions**, we mentioned that they can be deployed with a **trigger** that runs them when a file is uploaded, modified, or deleted from a specific bucket. This is a very convenient way to automate file management using an asynchronous serverless model where we only pay when there is activity in our buckets.

From a security point of view, apart from the already mentioned IAM integration, it's important to know that **server-side encryption** is used for our data by default, using keys managed by Google, with the option to use **customer-managed encryption keys** if we prefer.

When an object is uploaded to a bucket using the same name as an existing one, it will be overwritten. However, **object versioning** can be enabled to keep old versions, too. This will increase the number of stored objects and the associated cost, but it will also allow us to recover old versions of a file if needed, which can be a lifesaver depending on our requirements. I have seen a few customers struggle because of not enabling this feature early enough.

A **retention policy** can also be defined so that uploaded objects are deleted after a specific amount of time. This makes Cloud Storage useful for applications that need to follow data retention regulations for legal purposes or, for example, for services where old content is automatically deleted after some time to leave room for newer uploads.

Finally, Cloud Storage pricing is calculated by combining three different factors: the amount of data stored in our buckets, the processing that's done by Cloud Storage, and the amount of data read from or moved between our buckets. The free tier includes 5 GB of regional storage for free (in US regions only), together with 100 GB of network egress from North America to all region destinations (excluding China and Australia) per month.

Now, let's deep dive into the two key concepts of Cloud Storage.

Bucket locations and storage classes

When we create a bucket in one of our projects, we will be asked to choose a geographic location for it. This is an important choice because it can't be changed later. If we make a mistake, we can always create another bucket and move our files there, but given that bucket names are unique, this may not be an easy or comfortable transition in some cases.

We will be offered three choices for the location type of a new bucket:

- **Region**: Our objects will be stored in a single geographic place. This is the most affordable option.
- **Dual region**: Our objects will be stored in two geographic places. This is the most expensive option.
- **Multi-region**: Our object will be stored in a large geographic area with two or more geographic places. This option is cheaper than a dual region but more expensive than a single region.

Since each region has at least two available zones, as we mentioned in the previous chapter, all Cloud Storage data is always zone-redundant, and dual or multi-region types add region redundancy, with data replicated in areas located at least 100 miles away from each other.

Replication will happen in 1 hour or less by default, but we can enable *Turbo Replication* if this is not fast enough, and in that case, replication will happen in 15 minutes or less at an extra cost.

And if any of the regions become unavailable before an object has been replicated to them, Cloud Storage will make sure that stale versions of an object are never served, and objects will be replicated as soon as the region is available again. All of this is part of a feature called **strong consistency**.

And speaking of objects, when our applications use them to store files, there can be different requirements regarding retention and accessibility. For example, a backup log file may be required to be available for a long time but rarely be accessed, while images for a website or app may be used only for a few days, but will be frequently read by our visitors.

This takes us to the concept of a **storage class**, a piece of metadata attached to an object that defines the conditions under which that object will be stored, including information such as how long it will be stored and what its level of availability will be.

We can define a default storage class when we create a bucket, and Cloud Storage will attach that storage class to any object that is created inside that bucket. We can also change the storage class for one or more objects anytime we want.

These are the storage classes that we can attach to any object or set by default in a bucket:

- **Standard storage**: This is the best class for frequently accessed data and provides the highest level of availability. It doesn't have a minimum storage duration and no retrieval fees are incurred. It can be combined with dual- or multi-region storage options to increase redundancy and reduce latency.

- **Nearline storage**: This is a low-cost option that's ideal for files that need to be read or modified, on average, once a month or less frequently. This class is also useful for long-tail static assets. The minimum storage duration is 30 days and low-cost retrieval fees will be incurred.

- **Coldline storage**: This is a very low-cost option for storing data that is read or modified less than once a quarter. The minimum storage duration is 90 days and very low-cost retrieval fees will be incurred.

- **Archive storage**: This is the lowest-cost option for data access and happens less than once a year. This is the ideal option for long-term data archiving, backups, and disaster recovery data. Despite its slightly lower availability when compared with the Standard class, files will still be available within milliseconds. The minimum storage duration is 365 days and extremely low-cost retrieval fees will be incurred.

We just covered the basics of Cloud Storage. Next, let's change topics and introduce Cloud Tasks, an interesting service for repetitive tasks.

Cloud Tasks for asynchronous task execution

Our solutions or services will often need to deal with repetitive tasks, and it's in this kind of scenario that we can make the most out of our cloud provider. Processing millions of images or videos or making thousands of API calls to download reports can take a long time if we don't create a fast and robust architecture.

Having multiple workers for different tasks, implementing parallel processing, or being able to retry failed operations with an exponential backoff can complicate our code quite a lot and make it difficult to maintain.

Google Cloud offers different solutions for these kinds of challenges. We already covered some of them earlier in this book, such as Cloud Functions, App Engine, and Cloud Run, but this time, I want to include **Cloud Tasks** because, in my opinion, it's a very good example of a managed service that can save us a lot of time when we are creating a new application or migrating a legacy one.

Cloud Tasks is a managed service that allows us to create different **queues**, where we can **offload requests** to be processed asynchronously. We will map each of these queues to a worker that will complete the required task for a single request. Cloud Tasks will then take care of executing our worker multiple times in parallel to process all queued requests, giving us full control over how each queue will work: how many threads will run in parallel, how many times a failed operation will be retried, when to consider that a request timed out, and if and how to implement exponential backoff between retries.

Imagine a simple example: we need to create thumbnails for 3 million images every half an hour. There are different ways to implement this in Google Cloud, but Cloud Tasks is an interesting option. We can just create a queue and configure it to accept requests that include the URL of an image in Cloud Storage and the name of the bucket and path where the thumbnail must be stored.

Then, we create a Cloud Function or set up an App Engine service that takes these three parameters and creates a thumbnail for a single image. Finally, we need to create another Cloud Function that detects new files uploaded in our Cloud Storage input bucket and queues one request for each of them to Cloud Tasks; we will use Cloud Scheduler to run it every half an hour.

Just with these few steps, we have easily built a scalable system that will create our thumbnails as fast as possible and at a very low cost. And best of all, if tomorrow we need to crop these or other images too, we can reuse many of these components and implement the new service in a matter of minutes!

Cloud Tasks can help us move those tasks that are not **user-facing** to the background so that we can quickly answer any request. For example, imagine that a user requests complex processing of a video file using our application. Our code can just queue the processing request in Cloud Tasks and show the user a web response, almost immediately, saying that an email will be sent when the processing is completed, which improves the user experience.

The actual processing of the uploaded video will begin later, once the request reaches the top of the queue, and the user will be notified once the processing is complete, as the last action of the associated worker.

Cloud Tasks queues can be configured using the **gcloud** command-line tool, but for compatibility reasons, we can also use the AppEngine queue.yaml SDK file to create and configure our queues. The Cloud Tasks API is our third and last option, using either the REST API or the client libraries available for multiple programming languages.

Compared to **Cloud Pub/Sub**, Cloud Tasks provides scheduled delivery and task creation deduplication, together with individual task management, as differential features, apart from the fact that it is more targeted at explicit invocations, while Pub/Sub is often used for implicit invocations.

And while **Cloud Scheduler** acts more as a scheduler for repetitive tasks using a unique trigger, when we compare it with Cloud Tasks, we can see the latter more as an asynchronous task manager that supports multiple tasks and triggers.

Cloud Tasks is billed by every million billable operations, including API calls or push delivery attempts. The first million is offered at no cost every month, as part of the Free Tier. Services used or invoked using API requests are charged separately.

Next, we'll talk about Datastore, a flexible, fast, and scalable database.

Firestore in Datastore mode for high-performance NoSQL storage

Firestore in Datastore mode is a NoSQL fully-managed database that supports atomic transactions and provides flexible storage and massive scalability, with encryption at rest. It's a very interesting option for web backends and mobile applications.

Firestore in Datastore mode is a rebranding and the next evolution of Datastore. This mode combines Datastore's system behavior with Firestore's storage layer, which makes queries strongly consistent and removes previous limitations on the number of entity groups and writes per second, which were applicable when using Firestore in its legacy **Native mode**. On the other hand, when we use Firestore in Datastore mode, we need to use the Datastore viewer in the Google Cloud console, and the Datastore API and client libraries in our code. You can read more about the differences between the two Firestore modes on this page of the Google Cloud documentation site: `https://cloud.google.com/datastore/docs/firestore-or-datastore`.

Compared to traditional databases, basic concepts change their names:

- Tables become **kinds**
- Rows become **entities**
- Columns become **properties**
- Primary keys become **keys**

The main additional difference is that Firestore in Datastore mode is **schemaless**, which makes it much more flexible. For example, different entities (rows) in the same kind (table) can have different properties (columns), and these properties can use the same name but contain values of different types.

This makes Datastore especially useful in scenarios where different product types coexist, using different field names and types for each product. We no longer need hundreds of empty fields to support many different products at once – Datastore adapts to our needs instead of it being us who need to adapt to available features and capabilities.

Datastore supports **SQL-like queries** using GQL and **indexes**, together with keys, but it does not support **joins**, which may be a significant limitation in some use cases. As a fully **managed service**, Google takes care of data sharding and replication to ensure both **consistency** and **availability**.

Also, we can execute multiple datastore operations in a single Datastore transaction due to its compatibility with all four **ACID** characteristics: **atomicity, consistency, isolation, and durability**.

Datastore data can be accessed using its JSON API, open source clients, or community-maintained **object-relational mappings** (**ORMs**), such as Objectify or NDB.

The Free Tier of Datastore includes 1 GB storage, 50,000 entity reads, 20,000 entity writes, 20,000 entity deletes, 50,000 small operations, and 10 GiB of egress per month. Extra entity operations are paid per use in groups of 100,000, while additional egress is charged by GiB used each month.

Now, let's talk about Cloud Workflows and how it can help us connect services to build an end-to-end solution.

Cloud Workflows for service orchestration

When we build a solution using microservices or if we simply create different pieces that handle a part of the total job, we need a way to connect all these services and order them in a way that makes sense, where the outputs from one step will become the input to the next one.

For example, if we create an application to convert audio recordings recorded by our Customer Support department into a dashboard of sentiments, one of our components will detect the audio file format and route the file to the corresponding processor. Then, the processor will convert the audio file into a common audio format, another service will convert the audio file in the common format into text, the next one will analyze the language and create insights, and a final service will consolidate all the information, generate a dashboard, and email a sentiment report when it's ready.

Cloud Workflows can help us build solutions like the one I just described, integrating Google Cloud Public APIs such as natural language processing with custom services, such as our Cloud Functions, App Engine, or Cloud Run, and with any other HTTP-based APIs, internal or external.

This is a **managed service** and we also have all the benefits of a **serverless** solution – that is, on one side, we will not be charged when our workflows are idle and on another, there will be no infrastructure to patch or security to harden. We just care about designing our workflows and Cloud Workflows will do the rest for us.

There are many use cases for which this service can be a nice fit, including service **orchestration** to build solutions, handling repetitive jobs, automating IT delivery or business processes, and any other scenarios where an ordered execution of services can be of help.

Workflows are written using either **JSON** or **YAML** files, describing each of the steps that should be executed. Different blocks can be used in each step, including multiple **conditional** statements based on the value of a variable, **iterations** over a collection of data, or executing a step **concurrently** to process multiple values in parallel.

We can also group steps that repeat often into a **sub-workflow**, which acts similarly to a function written in any programming language, and access global parameters passed to the workflow in any of the steps. All these features will allow us to create rather complex workflows that can be used to solve many different challenges.

Once our workflow is ready, we can run it using the UI or the `gcloud` command-line utility, use the REST API or a client library to execute it from our code, or schedule it to run one or more times using **Cloud Scheduler**, a service that we will talk about at the end of this chapter.

The pricing of Workflows is calculated monthly based on the number of workflow steps executed, regardless of their result and whether they are first or subsequent retries. External API requests, resources with custom domains, and steps waiting for callbacks are considered **external steps**, while the rest are internal. The first 5,000 internal steps and 2,000 external HTTP calls per month are offered at no cost, as part of the Free Tier.

We will create a workflow in our *Hands-on exercise* section so that you understand how this service works and how useful it can be to connect all our microservices and turn them into actual solutions.

Speaking about service connection and communication, it's time to talk about Pub/Sub.

Pub/Sub for inter-service messaging

Pub/Sub is a messaging service that works well with distributed applications and those using microservice architectures. Its key advantage is that it provides an asynchronous service that decouples message producers from message consumers.

Inter-process communication has been traditionally performed using **remote procedure calls** (**RPCs**) and similar **synchronous** systems. Pub/Sub offers an alternative based on **asynchronous broadcasting**, where event producers send a global message attached to a topic, and event consumers subscribe only to those **topics** that are of interest to them.

Each of the messages is indeed *lent* to all active subscribers until they are acknowledged by one of the consumers, indicating that it has been processed, after which they are removed from the queue.

Since a message can contain all kinds of information, this makes it possible to send data across processes and services using a scalable and asynchronous system. Services sending messages can do so whenever and as many times as they want, and services developed to receive and process those messages can do so at a different time, without causing any disruptions or data loss in the communication process.

For example, imagine a highway toll system, where all daily toll charges are sent at night to a central processing system and sent for payment the next morning. We could use Pub/Sub for all toll stations to stream their daily pending charges at 11 P.M. and use a central receiver to start reading and processing those payments the next morning.

Pub/Sub is offered in two different versions:

- **Pub/Sub standard**: Replicates data to at least two zones, with a third added with a best-effort approach. No additional maintenance is required.
- **Pub/Sub Lite**: This is a lower-cost option where topics are available only within a single zone or region. If we choose this option, we will need to provide and manage our storage and throughput capacity, an additional load that may not justify this choice in most cases.

We can use Pub/Sub in a lot of different scenarios and for many different uses, such as these:

- **Parallel processing**: We can have a single event producer and multiple consumers that will process queued messages in parallel, maximizing the performance of our application.
- **Event ingestion**: We can have multiple event producers streaming a massive amount of events that we will be queued and that we will ingest asynchronously on our backend using one or more consumers, maybe at a different time.
- **Event distribution**: If we need to share global information to be used by multiple departments or applications, we can broadcast them using different topics and let each department subscribe, receive, and process only messages from those topics that they are interested in.
- **Data streaming or replication**: We can stream our data row by row or in slices and use a subscriber to replicate it in another location.

As with the rest of the services, we can interact with Pub/Sub using the UI, run commands using `gcloud`, or use either the REST API directly or via a client library for our favorite programming language.

The cost of this service is based on the combination of throughput, egress, and storage required for our messages. The first 10 GiB of throughput is free every month as part of the Free Tier.

Due to the nature of this server, using an emulator for testing is recommended. I added a link about this topic in the *Further reading* section.

Next, let's present another service that will help us protect our secrets.

Secret Manager for storing sensitive data

When we write our code, there is a lot of sensitive data that we use and should never be stored in a repository, such as **credentials**, **API keys**, **certificates**, or **database passwords**. Even if leaving this data out of our code is a good practice, where can we store it, access it, and easily update it when required without compromising security? This is where **Secret Manager** comes to the rescue.

Secret Manager is a managed service that allows us to store any kind of configuration information or sensitive data, stored either as **text strings** or **binary blobs**, and retrieve it any time we need it.

Each piece of sensitive data is stored as part of what is called a **secret**, which also contains additional **metadata**. **Versioning** is enabled for the contents of each secret, so we can always recover an old password if it's needed, or just get the latest version to ensure that we are always up to date. When we add a new version, the secret is said to be **rotated**. Besides, any specific version of a secret can be **disabled** if we need to prevent access to its contents.

Using the Secret Manager API directly is the recommended implementation path to maximize security. Remember that security is based on the **principle of the weakest link**, so practices such as retrieving a secret just to store it in a file or passing it using an environment variable can put your entire application at risk.

Also, rotating secrets periodically will mitigate the risk of leaked data, which will make them unusable after the rotation. Configuring IAM wisely and enabling **data access logging** can also help us troubleshoot and identify the root cause of any security issues.

This service can also be used in combination with **Cloud KMS** to create cryptographic keys and use them to encrypt or decrypt our secrets.

Secret Manager is billed depending on the number of active secret versions per location, the number of access operations, and the number of rotation notifications. The Free Tier includes 6 active secret versions, 3 rotation notifications, and 10,000 access operations every month.

In the following section, we'll discuss how we can schedule jobs and tasks in Google Cloud.

Cloud Scheduler for running jobs at the right time

This is the service offered by Google Cloud for running tasks at specific times. We can schedule a Cloud Function to generate and send a daily report at 7 A.M. from Monday to Friday, or we can invoke our App Engine service to process pending user-uploaded images every 10 minutes. If you are a Linux or Unix user, this is the equivalent of **cron**.

Cloud Scheduler can schedule units of work to run periodically and supports the following types of targets:

- A **HTTP(S) endpoint**
- A **Pub/Sub topic**
- An **App Engine HTTP(S)** service
- **Cloud Workflows**

For App Engine, the job must be created in the same region as the App Engine app, while for the other three options, any Google Cloud region where the service is available can be chosen.

Cloud Scheduler jobs can be created from the UI in the Cloud console, using the `gcloud` command-line utility, or using either the REST API or a client library. The UI will use a second tab in the Cloud Scheduler interface to show information about jobs scheduled in App Engine using `cron.yaml`.

For example, creating a job using `gcloud` would be as simple as this:

```
gcloud scheduler jobs create http sample-http-job \
    --schedule "0 7 * * *" \
    --uri "http://mydomain.com/my-url" \
    --http-method GET
```

This would schedule a job every day at 7 P.M., invoking the specific URL included previously. The scheduling time is defined using the cron format. I added a link in the *Further reading* section at the end of this chapter to help you get familiarized with this format if you have never used it.

As shown in the preceding example, additional parameters can be used, depending on the target type, such as the HTTP method, the Pub/Sub topic, the Workflow name, the App Engine location, or relative URL. A job can also be configured to run as a specific service account to enable authentication using either **OAuth2** or **OpenID Connect** tokens. We will see an example in our hands-on exercise.

Other important parameters are the region where the job will run and the time zone that the schedule will use. Please take into account that if we don't choose a time zone carefully, our execution times may change during any **Daylight Saving Time (DST)** period. If this will be a problem for our service, we should choose a UTC-based time zone.

Finally, we can also configure a specific number of retries in case of errors, including a backoff strategy to increase the interval between failed jobs, which can help us support both instant failures and also longer-running ones. For these strategies to work, it is key that our application provides a proper **return code** or **status code** to indicate the result of the operation.

Cloud Scheduler is billed at a fixed cost by job ($0.10 at the time of writing this chapter) and all accounts get 3 services as part of the Free Tier. Notice that this is at the account level and not at the project level.

Now, it's time to put all these services into action in our hands-on exercise.

A hands-on exercise

In this exercise, we will build a solution using most of the services covered in this chapter. Our objective is to provide a web page containing the current weather information for a list of cities. This can be implemented using different architectures. Let me describe one of them and then we can comment on what other options we could have chosen.

In this case, I identified the following list of steps:

1. Read the list of cities.
2. Get the forecast for each city.
3. Put the results in a centralized storage location.
4. Update the web page containing the weather information.
5. Repeat all previous steps periodically.

A lot of questions need to be answered before we can move on; these are some that came to my mind during the process and how I came to answers to them. You may disagree with some of them, and that's perfectly fine.

First, I decided to use an external list of cities stored in a file, containing just the name of each of them, so that anyone could make changes without needing updates in the code. I thought that a file in Cloud Storage could be a nice option since it's also one of the services described in this chapter. Additional options could be a Google Docs spreadsheet, a file in Google Drive, and more.

Then, I had to find an open API providing weather information. I chose **Weather API** (`https://weatherapi.com`) because they have a free tier that just requires a link back and they offer a nice API that provides the weather by city name.

For centralized storage, I chose Datastore because I can create a custom kind and store a single entry associated with each city name and assign it to the full response provided by the API, which I can use later. Data updates and listing all entries are really easy operations, too.

For the final web page, I could have set up a Cloud Function or an App Engine service but decided to use a public Cloud Storage bucket where I could directly serve the content so that you could understand how to set it up and how to mitigate potential risks.

Last, but not least, I decided to orchestrate the whole process using Cloud Workflows to show you how simple and useful it can be. And, as we discussed, it can be integrated with Cloud Scheduler, so our idea of updating the web page periodically can be easily implemented too.

Let's begin by cloning the source code from the repository for this book in our Cloud Shell console. All the required files are in the subdirectory for this chapter. Also, either create a new project or choose one that you created before, find its project ID, and don't forget to configure it by running the following command:

```
gcloud config set project <YOUR_PROJECT_ID>
```

Now, it's time to create our first Cloud Function.

Reading the list of cities

This was the first step I identified. I decided to create a Cloud Function that reads a file from Cloud Storage and returns the list of city names in JSON. My input file for tests, named `city_list.csv`, looks like this:

```
Madrid
london
toKyo
HousToN
lima
Pretoria
```

As you can see, I mixed upper and lowercase letters to check how well the API behaved in all these different cases, and I must admit that it worked well.

I created a private Cloud Storage bucket called `sampleprivate384` in a single location (*us-east1*). Using the Standard storage class, I uploaded my file in a folder called `input`, so the full path to the CSV file containing the list of the cities is `input/city_list.csv`. Please proceed to do the same in your project; just choose a different name and write it down.

This first Cloud Function is called `weatherinfo-readcitynames`. I recommend using a common prefix in the name so that all the resources belonging to a specific solution are easier to find.

The code for this function is located in the `readcitynames` directory in the repository for this chapter. The code will just take the name of the bucket and the path to the list file as parameters and will read all the lines and return a Python list containing the cities. Since we are using *Flask* to handle the request and the response, it will be automatically converted into a JSON structure when returned, and we can integrate it later in our workflow. This is an extract of the main code, where exception handling and comments were removed:

```python
storage_client = storage.Client()
bucket = storage_client.bucket(bucket_name)
blob = bucket.blob(city_list_path)
with blob.open("r") as city_list_file:
    city_list_raw = city_list_file.readlines()
city_list = []
for line in city_list_raw:
  city_list.append(line.strip())
return city_list
```

Also, notice how parameters are passed using JSON. This is the way Cloud Workflows works:

```
bucket_name = request.json.get('bucket_name')
city_list_path = request.json.get('city_list_path')
print(f'Reading list {city_list_path} in {bucket_name}')
return(get_city_list(bucket_name, city_list_path))
```

I also included a main() function in these and all other Cloud Functions so that the code can be run, for testing purposes, from the command line. I also added a few logging lines so that we can track the execution and verify that all the parameters work as expected. The list of cities is printed too before it's returned.

You can deploy this Cloud Function by running the following command from the readcitynames directory. You may prefer a different region; choose the one you prefer and use it for these and the rest of the deployments in this exercise:

```
gcloud functions deploy weatherinfo-readcitynames \
--gen2 --runtime=python310 --region=us-east1 \
--memory=256MB --source=. \
--entry-point=read_city_names_trigger \
--trigger-http --allow-unauthenticated
```

Next, it's time to get the weather information.

Getting weather information for each city

The second Cloud Function will retrieve the weather information for a specific city and will return the raw API response. To use the API, we will need to sign up and get an API key. Let's do so by going to https://www.weatherapi.com/signup.aspx.

Once you fill in the form, check your email; you will receive a message with an account activation link that you should click on. Then, just log in and copy your API key.

Since this API key is a sensitive resource and could be changed at any time, I decided to store it in Google Secret Manager. To do this, you can open the product section in the Cloud console at https://console.cloud.google.com/security/secret-manager or just search for Secret in the omnisearch box at the top of the Cloud console.

Once there, click on + **Create Secret** and use weatherapi_key as the name, and paste your API key into the Secret Value text box. Our API key is now securely stored, and we will use the Secret Manager API to retrieve it in our code.

In terms of parameters for this Cloud Function, we will just need the name of the city and the Project ID, which Secret Manager will require to retrieve the secret.

I decided to use an optional `api_key` parameter so that I can hardcode an API key during my tests. If this parameter is not passed, it becomes None by default and the secret is retrieved. This is a summary of the code for retrieving the secret:

```
if not api_key:
    # Create the Secret Manager client.
    client = secretmanager.SecretManagerServiceClient()
    # Build the resource name of the latest version of our secret to retrieve.
    secret_name = f'projects/{project_id}/secrets/weatherapi_key/versions/latest'
    # Retrieve the value of the secret
    result = client.access_secret_version(request={'name': secret_name})
    api_key = result.payload.data
```

Figure 9.1 – Code for retrieving the value of a secret from Secret Manager

Notice that the name of the secret is hardcoded, as it is the *latest*, which is an alias to always retrieve the up-to-date version of a secret.

Now, we can use Weather API and return the result with just a few lines of code. This is a summary:

```
URL = "http://api.weatherapi.com/v1/current.json"
PARAMS = {'q': city_name, 'aqi': 'no', 'key': api_key }
api_result = requests.get(url = URL, params = PARAMS)
return api_result.json()
```

Notice how we pass the city name as a parameter, together with the API key. The `aqi` parameter is used to include **Air Quality Information (AQI)** in the response. I disabled it because I didn't want to use it, but I included a sample API response for *London*, including AQI, in the repository for this chapter, in a directory called `sample_api_response`.

You can deploy this Cloud Function using a command like the earlier one:

```
gcloud functions deploy weatherinfo-getcityweather \
--gen2 --runtime=python310 \
--region=us-east1 --memory=256MB --source=. \
--entry-point=get_city_weather_trigger \
--trigger-http --allow-unauthenticated
```

The next Cloud Function will store the response in Datastore.

Storing weather information in a central location

This Cloud Function will just receive a city name, a city weather information JSON response, and the Project ID as parameters and will store that information in Datastore. This part of the functionality could have been included in the previous Cloud Function, but I opted to keep it apart so that it can be easily updated if we ever decide to use a different storage location. Besides, I prefer each function to do a single task where possible, following a microservice approach.

This is a simple function whose simplified core code looks like this:

```
datastore_client = datastore.Client(project=project_id)
kind = 'city_weather'
details_key = datastore_client.key(kind, city_name)
weather_details = datastore.Entity(key=details_key)
weather_details['weather_details'] = city_weather_details
datastore_client.put(weather_details)
return "Sucess"
```

I used a kind called `city_weather` to store this type of information; then, I created a new entity associated with this kind and the city name and used it to store the weather details. This way, I can later request all entities of this same kind to build the final web page.

The following command is required to deploy this Cloud Function, similar to others seen before:

```
gcloud functions deploy weatherinfo-storecityweather \
--gen2 --runtime=python310 \
--region=us-east1 --memory=256MB --source=. \
--entry-point=store_city_weather_trigger \
--trigger-http --allow-unauthenticated
```

The final Cloud Function will update our HTML file containing the weather information for all the cities in our list.

Updating the weather web page

This is the last of the Cloud Functions, where we will generate and store the public web page. As input parameters, we will just need the Project ID (for Datastore), together with the name of a bucket and the full path to our HTML target file.

You can use the same bucket where the CSV file is stored, so long as it doesn't have uniform access enabled and public access is not prevented. I created a new bucket without these limitations and called it `samplepublic134`. This is a key part of the process. The following screenshot shows the options you can choose from when creating this bucket:

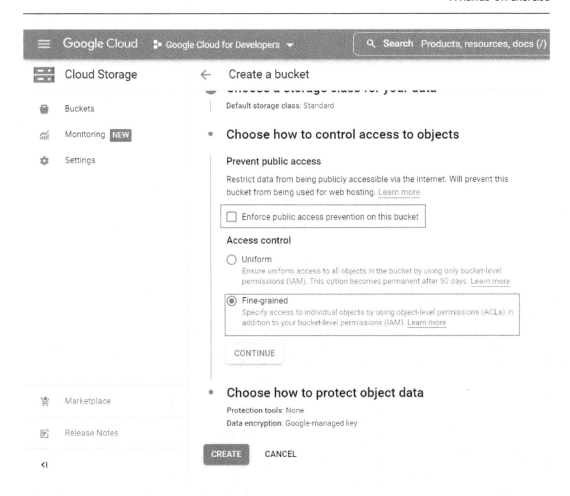

Figure 9.2 – Options to take into account when creating our public bucket

Now, we must retrieve all the entities of the `city_weather` kind and use parts of the weather information stored for each of them to build our web page, which we will then send to Cloud Storage. This is a summary of that part of the code:

```
datastore_client = datastore.Client(project=project_id)
query = datastore_client.query(kind='city_weather')
weatherdata_list = query.fetch()
...
storage_client = storage.Client()
bucket = storage_client.bucket(bucket_name)
blob = bucket.blob(html_path)
with blob.open("w") as f:
```

```
    f.write(html_content)
blob.content_type = 'text/html'
blob.cache_control = 'no-store'
blob.patch()
blob.make_public()
print(f'HTML file publicly accessible at {blob.public_url}')
return 'Success'
```

In the preceding code, after writing the file contents to the blob, I updated the content type and the cache control metadata using the `patch()` function to make sure, first, that the file is displayed in a browser instead of being sent as a download. Second, I wanted to ensure that the file is not cached so that we can get updated content as soon as it is available.

I am also using the `make_public()` method to make the file available to any user, turning Google Cloud Storage into a static web server for this specific file. Please remember that this function will fail to work if you create the public GCS bucket with uniform access control since in that case, the configuration cannot be applied at the file level and must be applied at the bucket level. Please use a bucket with fine-grained access control instead.

This is the deployment command that you can use for this Cloud Function:

```
gcloud functions deploy weatherinfo-updateweatherpage \
--gen2 --runtime=python310 \
--region=us-east1 --memory=256MB --source=. \
--entry-point=update_weather_page_trigger \
--trigger-http --allow-unauthenticated
```

Now, it's time to connect all these functions using Cloud Workflows.

The end-to-end workflow for our weather solution

Once we have built all the pieces, we can use Cloud Workflows to define the order of the steps to follow.

First of all, we will need a service account to run both our workflow and to invoke it later using Cloud Scheduler. Let's go to the **Service Account** area of IAM in the Cloud console by opening the following address in a new tab of our browser: `https://console.cloud.google.com/iam-admin/serviceaccounts`. If required, click on your Google Cloud project name to get to the account list screen.

Next, click on + **Create Service Account** and choose a name for the service account – for example, `weather-agent`. Then, click on **Create and Continue** and grant the Service Account the **Cloud Datastore User** role, click on **Continue**, and click on **Done** to finish creating the account. Remember the name of the account you just created, since you will need it very shortly.

Open the UI in the Cloud Console by either using the omnisearch box at the top to find `Workflows` or using the following link: `https://console.cloud.google.com/workflows`.

Click on **+ CREATE** and enter a name for the workflow (in my example, I used `update_weather_for_cities`). Select the same region as for your Cloud Functions and the service account you just created and click **Next**.

On the new screen, open the workflow contents provided in the `workflow` directory and paste it into the big text box. Notice how a few seconds later, a graph like the one shown in the following screenshot will be shown on the right-hand side of your screen, visually representing all the different steps in the workflow and how they are connected:

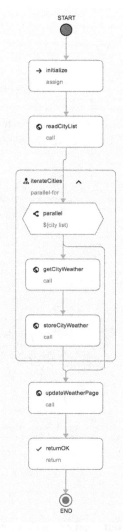

Figure 9.3 – Detail of the graph in the workflow editing screen

This graph summarizes how our workflow will work and is consistent with our idea of listing the cities, then iterating to get and store the weather information for each of those cities, and finally generating a web page containing all the information.

> **Important note**
>
> Notice that the workflow will process and store the weather information of all cities in parallel (notice the boxed part of the graph in the step called **iterateCities**) so that the execution will be faster. Once all cities are done, we will update the weather page with the information for all cities. This is an interesting option for this service.

Click on the **Deploy** button and wait for a few seconds; the workflow will be ready to be executed. Click on the **Execute** blue text at the top, then click on the blue **Execute** button on the next screen, and wait for a bit – you should see the workflow reaching the final step, `returnOK`, with a status code of `Succeeded`.

Now, if you visit the properties page of your public GCS bucket in the Cloud Console, you should see a file named `weather_by_city.html` available in the root folder. If you copy the public URL using the link located at the right-hand side of the row and paste it into your browser, that will be the URL where your weather information will be updated. The file looks like this on my browser:

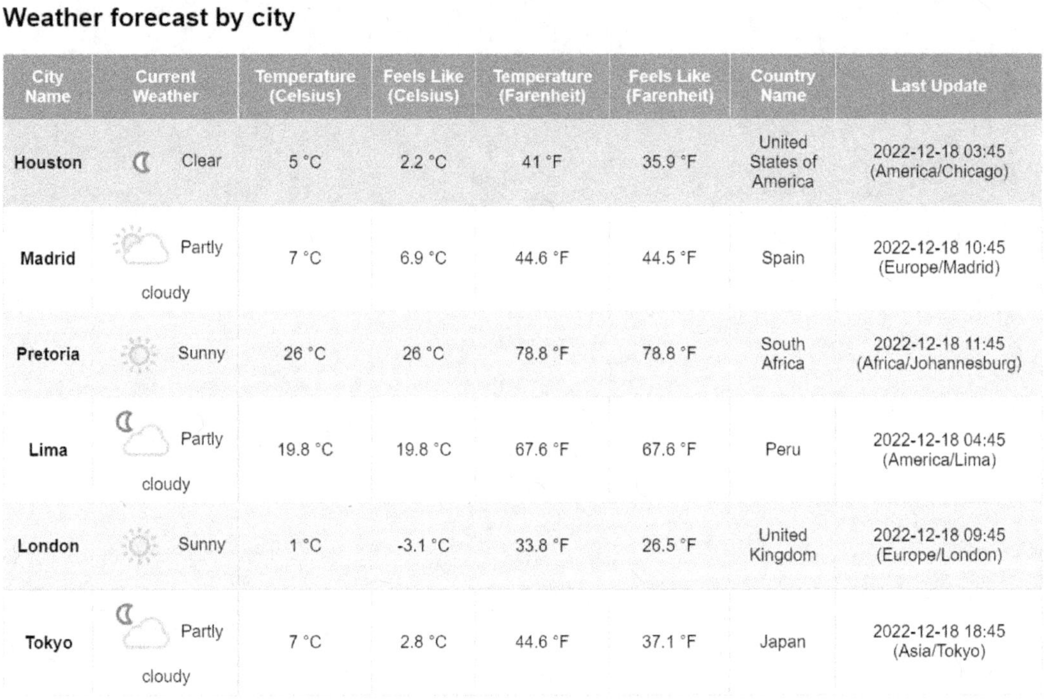

Weather forecast by city

City Name	Current Weather	Temperature (Celsius)	Feels Like (Celsius)	Temperature (Farenheit)	Feels Like (Farenheit)	Country Name	Last Update
Houston	Clear	5 °C	2.2 °C	41 °F	35.9 °F	United States of America	2022-12-18 03:45 (America/Chicago)
Madrid	Partly cloudy	7 °C	6.9 °C	44.6 °F	44.5 °F	Spain	2022-12-18 10:45 (Europe/Madrid)
Pretoria	Sunny	26 °C	26 °C	78.8 °F	78.8 °F	South Africa	2022-12-18 11:45 (Africa/Johannesburg)
Lima	Partly cloudy	19.8 °C	19.8 °C	67.6 °F	67.6 °F	Peru	2022-12-18 04:45 (America/Lima)
London	Sunny	1 °C	-3.1 °C	33.8 °F	26.5 °F	United Kingdom	2022-12-18 09:45 (Europe/London)
Tokyo	Partly cloudy	7 °C	2.8 °C	44.6 °F	37.1 °F	Japan	2022-12-18 18:45 (Asia/Tokyo)

Figure 9.4 – Weather information web page preview on a browser

Notice how I used the city name returned by the API instead of using the provided ones where upper and lowercase letters were used. I also used the icons provided by Weather API. Please remember that if you use this for any purposes beyond this exercise, you should add a link back to their website.

Now, we just need to schedule the workflow periodically and we will be done.

Updating our web page every 30 minutes

Since Cloud Scheduler is compatible with Cloud Workflows, this will be very easy to set up. Open Cloud Scheduler by searching for it in the omni box or opening its direct link: `https://console.cloud.google.com/cloudscheduler`.

Click on + **Create Job** at the top and choose a name for the scheduler (for example, `update_weather`), choose the same region as for the rest of the resources, and schedule it to run every 30 minutes using the `*/30 * * * *` string for frequency. Choose your time zone, even though it won't be relevant in this case because we are running the job at regular intervals. Click on **Continue**.

Next, select **Workflows via HTTP** for **Target type** and select both the workflow and the service account created earlier. Add the `roles/workflows.invoker` role to the service account if required either using IAM or by running the following command from the console, replacing the placeholders with your actual Project ID and Service Account address:

```
gcloud projects add-iam-policy-binding <PROJECT_ID> \
--member=serviceAccount:<SERVICE_ACCOUNT> \
--role=roles/workflows.invoker
```

Click on **CONTINUE** and look at the retry configuration options so that you can get familiarized with them, even though there is no need to change anything at this time. Then, just click on **CREATE** – our workflow will be scheduled to run every 30 minutes.

You can also run it manually from the main Cloud Scheduler screen, and use logging to verify it, or just reload our weather page, where I added a last update column to the right so that it's easier to see how recent the last update was. I included cities from all continents so that you can verify it more easily.

This is the end of our exercise for this chapter. I hope you found it interesting and that it helped you understand how you can use Google Cloud services to build better solutions faster.

But if you haven't had enough, let me share some tips, questions, and additional work ideas in the following section.

What's next?

Now that we are done with our exercises, here are some tips and some food for thought:

- All the deployment scripts use 256 MB of memory and allow unauthenticated invocations. How would you check whether more or less memory is required for any of them? For this specific case, does it make any sense to allow unauthenticated invocations?

- What happens if we remove a city from our list? How can we fix it or mitigate it? Are there any other weak spots that you can find in the architecture, the workflow, or the code? How would you improve or fix them?

- Some of the actions (retrieve secret, store weather information, read city list, and so on) can be directly implemented as workflow steps replacing cloud functions with REST API calls. Will you be able to make any of them work?

- Most of the actions that we performed using the console can be also implemented using `gcloud` commands – for example, creating the scheduler. Can you find the alternative commands for all these cases?

- What alternative implementations can you think of for this specific example, such as using Cloud Tasks, App Engine, Cloud SQL, and so on?

- Can you think of a similar example for working with other kinds of information, such as getting sports results, stock market updates, the latest news, release dates for books, movies, TV shows, and so on? Find a topic that interests you and try to adapt this workflow to implement it. You will realize how easy it is once you know the parts that you can combine to make things faster and more reliable. This is the magic of Google Cloud!

And now, it's time to wrap up.

Summary

In this chapter, we covered some of the basic Google Cloud services that we can use to simplify our development and migrations to the cloud.

First, we discussed how we can use Cloud Storage to store our files. Then, we introduced Cloud Tasks as a managed service for asynchronous task execution. Next, we talked about how Firestore in Datastore mode can help us store our data in a NoSQL database, with high performance and automatic scaling.

Next, we saw how Cloud Workflows is a great tool for combining multiple pieces and creating an end-to-end solution. Then, we went through the different uses of Pub/Sub for inter-component communication. In the next section, we talked about Secret Manager and how we can use it to store our most sensitive data. Finally, Cloud Scheduler was presented as a great option to ensure that our tasks and workflows run exactly when we want.

After covering all these services, we worked together on an exercise where we used them to provide weather information for multiple cities.

I hope you found this chapter interesting. In the next one, we will continue exploring Google Cloud services, but this time, we will cover those that are based on powerful machine learning models, offered using a public API, and that we can easily integrate to supercharge our application with advanced capabilities.

Further reading

To learn more about the topics that were covered in this chapter, take a look at the following resources:

- Manage encryption keys on Google Cloud with KMS: `https://cloud.google.com/kms`
- Datastore GQL reference: `https://cloud.google.com/datastore/docs/reference/gql_reference`
- Running the Datastore emulator: `https://cloud.google.com/datastore/docs/tools/datastore-emulator`
- Cloud Workflows code samples: `https://cloud.google.com/workflows/docs/samples`
- Testing apps locally using the Pub/Sub emulator: `https://cloud.google.com/pubsub/docs/emulator`
- Configure cron job schedules: `https://cloud.google.com/scheduler/docs/configuring/cron-job-schedules`

10

Extending Applications with Google Cloud Machine Learning APIs

The previous chapter covered Google Services that we can use as building blocks for our applications and services to make our development faster. Another way of achieving this purpose is using public APIs to integrate advanced services, providing features that we wouldn't be able to develop otherwise due to limitations in our budget, time, or even our knowledge and skills.

Google Cloud provides some remarkably interesting machine learning models that have been trained with huge amounts of data and can help us convert our unstructured data into structured data, a topic we will cover at the beginning of this chapter.

These models can be easily integrated into our code using their corresponding **Application Programming Interface** (**API**) and will add useful features to our applications, such as audio-to-text conversion, text translation, or insight extraction from text, images, or videos.

If you are interested in this area, you can also find a link to **AutoML** in the *Further reading* section. This more advanced product will allow you to create custom models for most of the AI solutions available in Google Cloud.

We'll cover the following main topics in this chapter:

- Unstructured versus structured data
- Speech to text
- Cloud translation
- Cloud natural language
- Cloud vision

- Cloud video intelligence
- Hands-on exercise

Unstructured versus structured data

As developers, we use different input and output methods and formats in our applications. One of the most common ones is **files**. We create files, copy, or move them to different locations or process them by reading and making changes to their contents.

A file contains, by default, what is known as **unstructured data**. This means that an audio, video, or text file is just data in a specific format. We can know the size of the file, what format was used to store the data, and maybe have access to some additional structured **metadata**, such as the creation date or the owner of the file, but we don't know anything about the contents. What's this video about? Is that audio file a song or a voice recording? Which language does this audio use? Is that text file a poem, or does it contain the transcription of a movie? Being able to answer these questions can enable more useful and impactful features in our applications.

The simplest example would be text files. Imagine we have a text file named `readme.txt` in Cloud Storage. From its metadata, we can know when it was created or modified and how big it is. We can even go one step further and read it using our code, counting the number of sentences, carriage returns, or punctuation symbols. While this can give us interesting information about the text, we still won't know which language it uses, what it is about, or even how the writer felt when it was written.

Machine learning (ML) and **artificial intelligence (AI)** can help us analyze and understand the content of our files, and Google Cloud offers different products in this area. For example, we can use **Translation AI** to translate all our texts into a common language. We can even use **Speech-to-Text** to convert audio files into text and obtain more input data. Finally, we can use **Cloud Natural Language** to analyze, classify, and annotate each of the files and extract interesting insights from our text files.

Are our customers happy with the product we just released? To answer this question, we can retrieve their public social media posts and perform sentiment analysis. We can also find out which of our customers reported bad experiences to our support team on the phone using Speech-to-Text. Or we can now understand what the best and worst valued areas in our business are by extracting names and adjectives from user surveys.

This can be even more powerful as we add media to the equation. We can use **Cloud Vision** and **Cloud Video Intelligence** to understand what's being shown on an image or when a specific object appears in a video. We can now understand why our users only watched half of our video or identify which characters keep them engaged. We can identify objects and colors, read and extract text, or detect different scenes and create structured metadata to classify our media and provide, for example, a much better searching experience for our users. The opportunities are endless!

To show you how easy it is to integrate these technologies with our code, let's start by explaining how Speech-to-Text can help us increase the amount of data we can analyze.

Speech-to-Text

When I discuss the available sources of information with customers, since breaking information silos is one of my priorities at Google, they often leave audio and speech data out of the picture, even if they usually have digital call centers with hours of support conversations recorded.

Adding speech as an additional source of information can help us access more data and obtain better insights by knowing our customers better. Besides, text is much easier to analyze than audio.

Speech-to-Text can help us **transcribe** audio to text and offers some useful features, such as specific enhanced models for phone calls and videos, multiple speaker labeling and splitting, automatic language detection, or word-level confidence scoring. Multiple audio formats are supported, such as WAV, MP3, FLAC, AMR, OGG, or WEBM.

Since the length of a sound file can vary, transcriptions can be performed **synchronously** or **asynchronously**. In the first case, our code will wait for the process to end before resuming the execution, while in the second, we will be returned an operation name that we can use to poll for status updates until the transcription is available. We can only use synchronous requests for local audio files lasting 60 seconds or less; otherwise, we will need to upload our file to Cloud Storage and pass the URI of the file instead of the file contents.

The following request can be sent to the API endpoint to synchronously transcribe a short standard WAV audio file recorded at 44 kHz from Cloud Storage:

```
{
    "config": {
        "encoding": "LINEAR16",
        "sampleRateHertz": 44100,
        "languageCode": "en-US",
    },
    "audio": {
        "uri": "gs://your-bucket-name/path_to_audio_file"
    }
}
```

I had some trouble sending this request using the *Try this method* section at https://cloud.google.com/speech-to-text/docs/reference/rest/v1/speech/longrunningrecognize, and as OAuth 2.0 is more flexible, I decided to use the OAuth 2.0 Playground instead, located at https://developers.google.com/oauthplayground/. This is a very useful tool for testing the GET and POST requests to Google API endpoints.

I used it to request a long-running operation to transcribe the audio and obtain an operation name. Then I used `https://cloud.google.com/speech-to-text/docs/reference/rest/v1/operations/get` as the endpoint to get the transcription after passing the previously obtained operation name as a parameter.

This is an excerpt of the result, which I included in the repository for this chapter:

```
{
    resultEndTime": "150.730s",
    "languageCode": "en-us",
    "alternatives": [
      {
        "confidence": 0.93837625,
        "transcript": " nasal is released through the nose"
      }
    ]
}
```

As you can see, each portion of the audio is transcribed into a block, for which one or more transcribing alternatives are provided, together with their confidence score, which is used to order them in the request from higher to lower.

The same request can also be implemented in Python using the client library. In this sample snippet, the first alternative for each block is printed for the sake of simplicity since they are ordered by confidence:

```python
from google.cloud import speech
client = speech.SpeechClient()
gcs_uri = "gs://your-bucket-name/path_to_audio_file"
audio = speech.RecognitionAudio(uri=gcs_uri)
config = speech.RecognitionConfig(
    encoding=speech.RecognitionConfig.AudioEncoding.LINEAR16,
    sample_rate_hertz=44100,
    language_code="en-US",
)
operation = client.long_running_recognize(config=config,
audio=audio)
print("Operation in progress, please wait...")
response = operation.result(timeout=90)
for result in response.results:
```

```
  print("Transcript: {}".format(result.alternatives[0].
transcript))
  print("Confidence: {}".format(result.alternatives[0].
confidence))
```

> **Important note**
>
> Notice that using service accounts is the recommended implementation option for the APIs included in this chapter for the sake of security. Loading a key or running the code as a Cloud Function deployed to run as a service account will help us mitigate potential risks, which may also affect our budget if a malicious user finds a way to make millions of calls.

Regarding pricing, the first 60 minutes of Speech-to-Text are free every month, and the rest are charged at $0.024 per minute. There is an option to log your data to help improve the quality of the ML models, and if you opt-in, the price is reduced to $0.016 per minute.

Transcribing is easy and can provide us with more text files to analyze. As a next step, we may want to translate all these texts into a common language before running a language analysis. In the next section, we'll see how we can use Translation AI for this purpose.

Cloud Translation

Translating text programmatically allows us to offer our services to users speaking different languages. **Cloud Translation** can help us understand their opinions and comments and even make them able to communicate with each other, thanks to its support for more than 100 language pairs. You can find a link to the full list of supported languages in the *Further reading* section, located at the end of this chapter.

Cloud Translation is a part of **Translation AI** and is offered in two editions: *Basic* and *Advanced*. I will only cover the Basic edition in the rest of this section, but you can find a link to a full comparison of both editions at the end of the chapter.

We can use the Cloud Translation API to translate texts or documents, supporting file extensions such as TXT, HML, TSV, PDF, DOCX, XLSX, or PPTX.

A basic translation request will include the text to translate, the source and target languages, and the format of the input data. I chose a very famous quote by William Shakespeare that will be translated into my native language, Spanish:

```
{
  "q": " To be or not to be, that is the question.",
  "source": "en",
  "target": "es",
```

```
    "format": "text"
}
```

The raw response from the API looks like the following. Notice the special character in the translated text:

```
{
   "data": {
     "translations": [
        {
           "translatedText": "Ser o no ser, esa es la cuesti\
u00f3n."
        }
     ]
   }
}
```

If I omit the source parameter because, in many cases, I will not know in advance the language used in the text, the API will detect it and return it as part of the response, which can be very useful, too:

```
{
   "data": {
     "translations": [
        {
           "translatedText": "Ser o no ser, esa es la cuesti\
u00f3n.",
           "detectedSourceLanguage": "en"
        }
     ]
   }
}
```

Now it's time to implement the same request using the client libraries for Python. You may need to enable the Translation API in your cloud project before the code works. Here, we will also let the API detect the source language. Notice also the use of u in the print statements to support printing Unicode characters:

```
from google.cloud import translate_v2 as translate
translate_client = translate.Client()
text = "To be or not to be, that is the question".
result = translate_client.translate(text,
```

```
target_language="es")
print(u"Text: {}".format(result["input"]))
print(u"Translation: {}".format(result["translatedText"]))
print(u"Detected source language: {}".
format(result["detectedSourceLanguage"]))
```

This is the output of the code, where the special character is correctly printed:

```
Text: To be or not to be, that is the question.
Translation: Ser o no ser, esa es la cuestión.
Detected source language: en
```

The pricing model for the *Basic* edition of Cloud Translation is based on the number of text characters processed, including the first 500,000 characters for free every month, while the rest are charged at $20 per million characters.

I hope this section helped you understand that translating is also very easy using the ML models available in Google Cloud. Now that all our text files use the same language, what kind of information can we obtain from them? Let's find it out in the next section.

Cloud Natural Language

Cloud Natural Language can help us analyze and understand the contents of a text file. We may not even know which language that text is written in, and the Natural Language API will detect it for us and provide that information as part of its response.

> **Note**
>
> Please take into account that while the list is constantly growing, not all languages are supported for all functions. You can find an up-to-date list at the following URL: `https://cloud.google.com/natural-language/docs/languages`. I also added this address to the *Further reading* section, so you can always have it to hand.

Once our text is ready, there are five different types of analysis that we can run on all or part of the contents of a text file:

- **Content classification**: This analyzes the provided text and assigns it a content category.
- **Syntactic analysis**: This breaks up the text into sentences and tokens and provides additional details about each of them.
- **Entity analysis**: This detects known entities, such as landmarks or public figures, and returns information about each of them.

- **Sentiment analysis**: This analyzes the text from an emotional point of view, returning if the writer's attitude is positive, negative, or neutral.

- **Entity sentiment analysis**: This combines the former two, providing a list of entities together with the prevailing emotional opinion for each of them.

We can run each of these analyses individually or combine two or more of them to be run simultaneously using the **Annotate Text** request.

Let's try an example: I looked up the first summary sentence about *"Tolkien"* in Wikipedia and took the liberty of making some small changes to add my personal opinion, resulting in this paragraph:

"John Ronald Reuel Tolkien was an English writer and philologist. He was the author of The Lord of the Rings, an amazing fantasy book you will love to read."

This can be a nice example for entity analysis because it's a short text that combines the name of a famous writer with one of his most famous works and my own (positive) opinion.

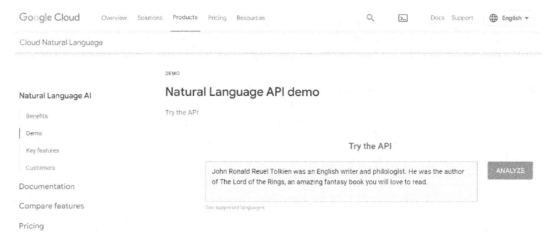

Figure 10.1 – Testing Cloud Natural Language AI using the demo website

Let's feed this text to Natural Language API demo available at the following URL: `https://cloud.google.com/natural-language`. Just go to the **Demo** section using the menu on the left side of the screen, paste the text as you can see in the preceding screenshot, click on the blue **Analyze** button and complete the Captcha, and you should see the results of the entity analysis looking similar to this:

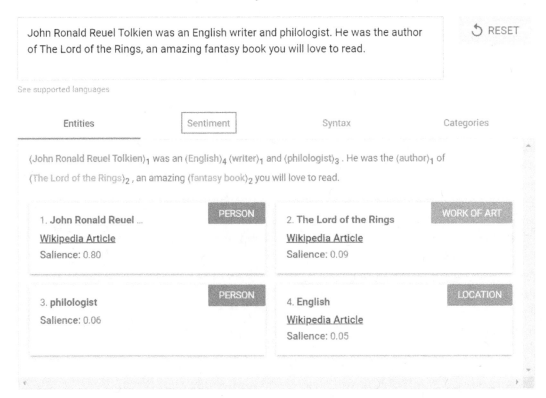

Figure 10.2 – Results of the entity analysis

As you can see in the results, the author's name, his place of birth, and his work are detected and linked to their corresponding articles in Wikipedia. A salience score is also included, indicating the relevance of each entity within the overall text.

If you change to the **Sentiment** tab (marked in red in the preceding screenshot), your results should be similar to the following:

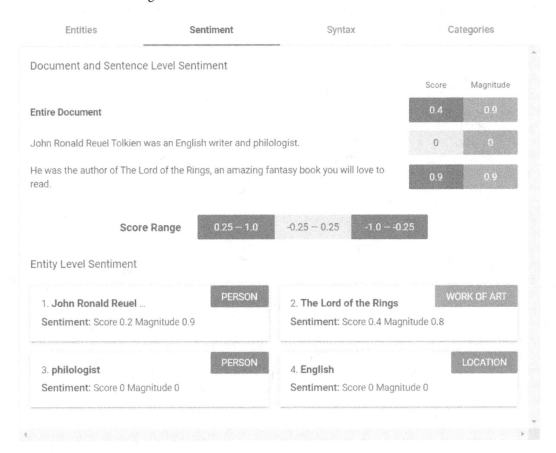

Figure 10.3 – Sentiment Analysis results

As you can see, **Score** and **Magnitude** are provided for each entity. The score is normalized between -1.0 (negative) and +1.0 (positive) and scores the overall emotion in the text, while the magnitude is not normalized and can be used to measure the absolute strength of the identified emotion.

Looking at the results, we can see how AI is able to understand that I'm speaking positively about the book, but this opinion is also affecting the score of its corresponding author. This kind of analysis can be useful for identifying not only emotional points of view but also who or what we are speaking about.

Running this kind of analysis from our code is also really easy. For example, the first entity analysis on our text could also be completed using the REST API and a simple request like this:

```
{
    "document":{
    "type":"PLAIN_TEXT",
    "language": "EN",
    "content":"John Ronald Reuel Tolkien was an English
               writer and philologist. He was the author
               of The Lord of the Rings, an amazing
               fantasy book you will love to read."
},
    "encodingType":"UTF8"
}
```

Notice that I included the language in the request, which I could omit and let the ML model identify for me. I could also pass the location of the text file within Cloud Storage instead of the text itself, which would lead to an even simpler request, such as the following:

```
{
  "document":{
    "type":"PLAIN_TEXT",
    "gcsContentUri":"gs://bucket-name/folder/file-name.txt"
  },
}
```

And this is an excerpt of the code that requests an entity analysis using the Python client library. We will see the complete code in the *Hands-on exercise* section. As you can see, it is also simple to integrate with our existing code, and iterating on the results is simple:

```
from google.cloud import language_v1 as language
def sample_analyze_entities(your_text):
    lang_client = language.LanguageServiceClient()
    lang_type = language.Document.Type.PLAIN_TEXT
    document = {"content": your_text, "type": lang_type}
    encoding_type = language.EncodingType.UTF8
    response = lang_client.analyze_entities(
      request={
        "document": document,
```

```
          "encoding_type": encoding_type
        }
    )
    for entity in response.entities:
      print("Representative name: {}".format(entity.name))
```

A couple of important points to take into consideration both for REST and Client Library implementations are included in the following list:

- Both HTML and TEXT files are supported; you can select your preferred format using the `Document Type` field.

- The `encoding_type` parameter is also important since the API calculates and returns offsets for each portion of the text that is analyzed. If this parameter is omitted, all returned offsets will be set to `-1`.

Now, let's take a look at the API responses. As you will see, the results for both examples are consistent with the ones we tried earlier on the demo site.

An excerpt of the response for our Entity Analysis request about Tolkien would look like this (the full response is available in the repository for this chapter):

```
{
  "entities": [
    {
      "name": "John Ronald Reuel Tolkien",
      "type": "PERSON",
      "metadata": {
        "mid": "/m/041h0",
        "wikipedia_url": https://en.wikipedia.org/wiki/
J._R._R._Tolkien
      },
      "salience": 0.79542243,
...
    {
      "name": "The Lord of the Rings",
      "type": "WORK_OF_ART",
      "metadata": {
        "wikipedia_url": "https://en.wikipedia.org/wiki/The_
Lord_of_the_Rings",
        "mid": "/m/07bz5"
```

```
      },
      "salience": 0.09017082,
...
    "language": "en"
}
```

Notice how the language document is returned at the end of the response and how each returned entity comes with a name, type, salience score, and metadata, including external references and a **machine-generated identifier** (**MID**).

The MID associates an entity with an entry in Google's **Knowledge Graph**, which is unique for all languages, so this can be used to link together words or names in different languages, referring to the same entity.

That's all for the entity search. Next, this is the excerpt of the response for a sentiment analysis on the same text about Tolkien:

```
{
    "documentSentiment": {
      "magnitude": 0.9,
      "score": 0.4
    },
    "language": "en",
    "sentences": [
      {
        "text": {
          "content": "John Ronald Reuel Tolkien was an English
writer and philologist.",
          "beginOffset": 0
        },
        "sentiment": {
          "magnitude": 0,
          "score": 0
        }
      },
      {
        "text": {
          "content": "He was the author of The Lord of the Rings,
an amazing fantasy book you will love to read.",
```

```
          "beginOffset": 65
      },
      "sentiment": {
        "magnitude": 0.9,
        "score": 0.9
      }
    }
  ]
}
```

In this response, notice first how each sentence comes with the corresponding offset, which works because the request included `encoding_type` set to `UTF8`. Also, see how the first sentence scores zero because there is no sentiment at all, but the second one, where I added my opinion, gets a 0.9 positive score and a decent magnitude level, even though I only included two words involving sentiments (but quite a powerful pair).

I hope these examples helped you understand the power of Cloud Natural Language. You will see a more detailed example, coded in Python, in the *Hands-on exercise* section.

The pricing model for Cloud Natural Language includes a free tier, and prices vary for the different analyses available. You can find all the details in the *Further reading* section.

Next, let's see what Google Cloud AI can do for us if we need to analyze images.

Cloud Vision

As I mentioned earlier in this chapter, we can get some information about a text file by counting characters or reading the size. In the case of images, the amount of information we can obtain is even less, mostly file size, image size, and maybe camera metadata.

However, as the adage says, "a picture is worth a thousand words," and there is a lot of information that we can get by looking at a picture. Cloud Vision, a product belonging to Google Cloud Vision AI, allows us to extract all this information in a structured way so we can *see inside pictures* and get some interesting information that we can use to obtain insights.

Let's give it a try first, so you can see what kind of information we can get. Let's try the demo again, this time available at the following URL: `https://cloud.google.com/vision`. Just go to the **Demo** section using the left menu.

This time we will use a picture of my good old Amstrad CPC 464, hosted by Wikipedia. This is the full URL to the image: `https://en.wikipedia.org/wiki/Amstrad_CPC#/media/File:Amstrad_CPC464.jpg`. Once you get to the **Media Viewer** page, click on the button with the arrow facing down to download the image (the following image shows you how to easily find that button):

Amstrad CPC 464, with CTM644 colour monitor

Bill Bertram - Own work   CC BY-SA 2.5

Figure 10.4 – Image of an Amstrad CPC with the download button highlighted in red

Then, just go back to the **Demo** page and drag the image to the **Try the API** section, or click that section to select it using the file browser. Next, just click on the Captcha test, and you should get to a results page that looks like this:

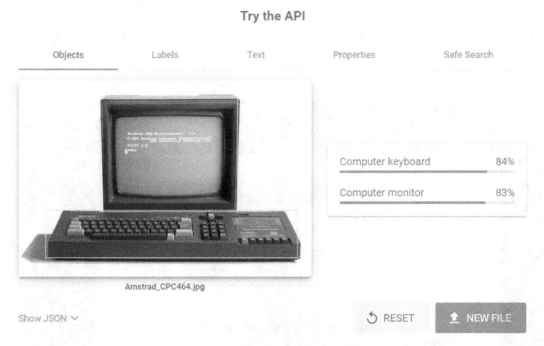

Figure 10.5 – Cloud Vision demo results

Notice how the content of the **Objects** tab mentions both the **monitor** and **keyboard**, returning the boundaries for each detected object and including a certainty score for each of them. This demo is more interesting (at least in my opinion) than the one for Natural Language because we can also click on the blue **Show JSON** banner at the bottom and see the raw API response. We will look at that response in a minute. I have also included a full sample response in the repository for this chapter.

Take some time to check the other tabs on the demo site. Notice how Cloud Vision also provides **Labels**, which describes elements detected in the picture, each of them also including its own **confidence score** and the coordinates of its **bounding box**. **Text** is also detected, and notice how AI can read the text on the Amstrad screen, despite its small size. The **Properties** tab also includes interesting information about colors and aspect ratio. And last but not least, the **Safe Search** tab helps us find how likely it is that an image includes **explicit content**, which may be problematic due to it being violent, racy, or containing adult content.

Now that we have already seen the practical side of Cloud Vision, let's formally enumerate what its API, **Cloud Vision API**, can do for us. There are quite a few different features available that, as for the Natural Language API, we can run either individually or combined into a single request:

- **Text detection**: This detects and extracts texts from an image, including bounding boxes and hierarchical information

- **Document text detection**: This is similar to the previous item but is optimized for pictures of documents with a dense presence of text

- **Landmark detection**: This identifies landmarks and provides a name for each one, together with a confidence score, a bounding box, and a set of coordinates

- **Logo detection**: This returns a name for each identified entity, a score, and a bounding box

- **Label detection**: This returns a set of generic labels with a description, together with a confidence score and topicality rating for each result

- **Image properties**: This includes a list of predominant colors, each with its associated RGBA values, confidence score, and the fraction of pixels it occupies

- **Object localization**: This detects multiple objects in an image, and, for each of them, the API returns a description, a score, and the normalized vertices for a bounding polygon

- **Crop hint detection**: This can be given up to 16 different aspect ratio values passed as parameters and provides a bounding polygon representing the recommended crop area, together with a confidence score and an importance score, representing how important the cropped area is when compared with the full original image

- **Web entities and pages**: This provides a list of resources related to the image, including entities from similar images, full and partially matching images, pages with matching images, similar images, and a best guess topical label

- **Explicit content detection**: This is also known as **SafeSearch** and provides six-degree likelihood ratings for explicit content categories, including *adult, spoof, medical, violence, and racy.*

- **Face detection**: This detects faces and their different elements, such as *eyes, ears, nose, or mouth,* with a bounding box and a score for each, together with a six-degree rating for emotions, such as *joy, anger or surprise,* and general image properties, indicating, for example, whether the image is underexposed or blurred

All these features open the door to amazing use cases that can help us make the most out of user-generated content and can also be used in many other scenarios where we can extract a lot of insights from our images.

You can read the full list of features in this section of the documentation website: `https://cloud.google.com/vision/docs/features-list`.

Continuing with our earlier example for the Amstrad CPC464 image, the following is the original REST request that was sent to the Cloud Vision API endpoint located at `https://vision.googleapis.com/v1/images:annotate`. Multiple features were requested in a single request to power the demo.

Notice how we can limit the maximum number of requests for each feature, select between different AI models in some features (check out the `DOCUMENT_TEXT_DETECTION` section for an example), and send different **aspect ratios** obtained by dividing the width by the height of an image to request crop hints:

```
{
  "requests": [
    {
      "features": [
        {
          "maxResults": 50,
          "type": "LANDMARK_DETECTION"
        },
        {
          "maxResults": 50,
          "type": "FACE_DETECTION"
        },
        {
          "maxResults": 50,
          "type": "OBJECT_LOCALIZATION"
        },
        {
          "maxResults": 50,
          "type": "LOGO_DETECTION"
        },
        {
          "maxResults": 50,
          "type": "LABEL_DETECTION"
        },
        {
          "maxResults": 50,
          "model": "builtin/latest",
          "type": "DOCUMENT_TEXT_DETECTION"
```

```json
      },
      {
        "maxResults": 50,
        "type": "SAFE_SEARCH_DETECTION"
      },
      {
        "maxResults": 50,
        "type": "IMAGE_PROPERTIES"
      },
      {
        "maxResults": 50,
        "type": "CROP_HINTS"
      }
    ],
    "image": {
      "content": "(data from Amstrad_CPC464.jpg)"
    },
    "imageContext": {
      "cropHintsParams": {
        "aspectRatios": [
          0.8,
          1,
          1.2
        ]
      }
    }
  }
  ]
}
```

Looking at the JSON response for the previous request, these are the sections included and the lines at which each of them starts:

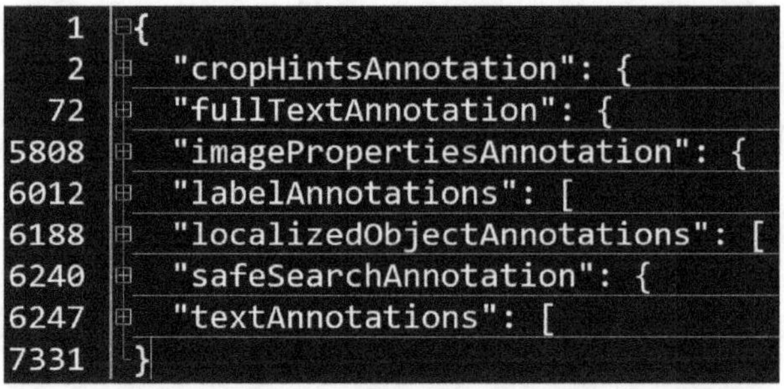

```
1   {
2       "cropHintsAnnotation": {
72      "fullTextAnnotation": {
5808    "imagePropertiesAnnotation": {
6012    "labelAnnotations": [
6188    "localizedObjectAnnotations": [
6240    "safeSearchAnnotation": {
6247    "textAnnotations": [
7331  }
```

Figure 10.6 – Sections and beginning lines in the Cloud Vision API response

Notice how the `textAnnotations` sections take almost 7,000 lines, while the rest just take a bit more than 300.

The layout of each result section is similar to what we saw for the Natural Language API. For example, as you can see in the following code block, one of the entries returned for the objects in the Amstrad CPC464 image. Notice the **normalized vertices** of the bounding box, the MID, the name, and the confidence score:

```
"localizedObjectAnnotations": [
    {
      "boundingPoly": {
        "normalizedVertices": [
          {
            "x": 0.09177671,
            "y": 0.66179305
          },
          {
            "x": 0.95162266,
            "y": 0.66179305
          },
          {
            "x": 0.95162266,
            "y": 0.9035502
          },
          {
            "x": 0.09177671,
```

```
                "y": 0.9035502
            }
        ]
    },
    "mid": "/m/01m2v",
    "name": "Computer keyboard",
    "score": 0.8360065
},
```

The entry for a label is simpler since there is no bounding box, but in this case, a **topicality score** is also included:

```
"labelAnnotations": [
    {
        "description": "Computer",
        "mid": "/m/01m3v",
        "score": 0.9721081,
        "topicality": 0.9721081
    },
```

This is another example of a part of the text seen on the computer screen. Notice how the detected text includes some noise and artifacts due to the small size of the letters. We should always choose the most useful results and filter them before we use them in other parts of our code. Finally, you can see how AI automatically detects the language of the text. This can also be useful in images containing text in different languages:

```
"textAnnotations": [
    {
        "boundingPoly": {
            "vertices": [
                {
                    "x": 213,
                    "y": 201
                },
                {
                    "x": 927,
                    "y": 201
                },
```

```
        {

          "x": 927,
          "y": 614
        },
        {
          "x": 213,
          "y": 614
        }
      ]
    },
    "description": "M\nAMSTRAD\n1802\n****\nW\nD\nC\nF\nG\
nAmstrad 64K Microcomputer (v1)\n$1984\nBASIC 1.0\nReady\nY\nH\
nand Locomotive Software L\nCPC 464 Coming\n98\nLA\nO\nP\nL\nJ
K L 7, G\nN, M. 3. 12\nCOOPUTOC",
    "locale": "en"
  },
```

Next, Safe Search confirms that this image is safe:

```
"safeSearchAnnotation": {
  "adult": "VERY_UNLIKELY",
  "medical": "VERY_UNLIKELY",
  "racy": "VERY_UNLIKELY",
  "spoof": "VERY_UNLIKELY",
  "violence": "VERY_UNLIKELY"
},
```

Finally, let's look at some code that uses the Python client library to detect objects in an image on Cloud Storage, with `uri` being the full path to the image to analyze. See how easy it is to invoke the API and iterate on the results:

```
def localize_objects_uri(uri):
    from google.cloud import vision
    client = vision.ImageAnnotatorClient()

    image = vision.Image()
    image.source.image_uri = uri
```

```
    objects = client.object_localization(
        image=image).localized_object_annotations

    print('Found {} objects'.format(len(objects)))

    for object_ in objects:
        print('\n{} (confidence: {})'.format(object_.name,
object_.score))
        print('Normalized bounding polygon vertices: ')
        for vertex in
          object_.bounding_poly.normalized_vertices:
            print(' - ({}, {})'.format(vertex.x, vertex.y))
```

You can find more information about the price of the different features offered by Cloud Vision in the *Further reading* section located at the end of this chapter.

Next, let's discuss how Video Intelligence API can help us extract a lot of information from our videos.

Cloud Video Intelligence

If Cloud Vision helps us extract information and insights from our images, **Video Intelligence** does the same with our videos. While some of their features are similar, others only make sense for videos. Let's enumerate them and add more information about the new features:

- Face detection.

- Text detection.

- Logo recognition.

- Label detection.

- Explicit content detection.

- **Object tracking**: This is similar to object detection in Cloud Vision, but the **Video Intelligence API** will also return **segments**, including an offset and a duration, to help us understand when the object was first seen in the segment and how long it was present.

- **Person detection**: This is a more advanced version of the face detection feature in Cloud Vision. It not only provides information about segments and bounding boxes but also provides details about upper and lower clothing color, sleeves, and the specific presence of body parts.

- **Shot change detection**: This feature detects abrupt changes in successive video frames and returns a list of segments.

- **Speech transcription**: This returns a block of text for each part of the transcribed audio. Currently, it only supports US English, with Speech-to-Text being an alternative available for other languages.

   This is also a very powerful tool for analyzing videos and extracting metadata and insights.

To run a test, I chose a free video by Vlada Karpovich, which you can download from the following URL: `https://www.pexels.com/video/a-person-pouring-water-on-a-cup-9968974/`. I uploaded this video to a Cloud Storage bucket.

Then, I used the API Explorer on the documentation page of the `videos.annotate` method, found at `https://cloud.google.com/video-intelligence/docs/reference/rest/v1/videos/annotate`. I used the right side of the screen to build the following request to perform label detection, object tracking, and explicit content detection on the formerly mentioned video. If you want to replicate it, you should replace the name of the bucket with your own:

```
{
   "features": [
     "LABEL_DETECTION",
     "EXPLICIT_CONTENT_DETECTION",
     "OBJECT_TRACKING"
   ],
   "inputUri": "gs://test_345xxa/pexels-vlada-karpovich-9968974.
mp4"
}
```

After clicking on the **Execute** button, I got the following response, which includes a name for the requested operation:

```
{
"name": "projects/292824132082/locations/us-east1/
operations/15179237131376606264"
}
```

This means that the processing of the video is happening in the background, and I have to poll the API until the results are ready.

To do that, I waited for a couple of minutes and then sent a post request to the endpoint for the `projects.locations.operations.get` method, available at the following URL: `https://cloud.google.com/video-intelligence/docs/reference/rest/v1/operations.projects.locations.operations/get`. I filled the parameter with the operation name obtained in the previous call and finally received the result of the operation in a JSON file containing 7,481 lines. I also included this file in the repository for this chapter. Let's look at the distribution of the contents of this file. As you will see in the following screenshot, object annotations take most of this file:

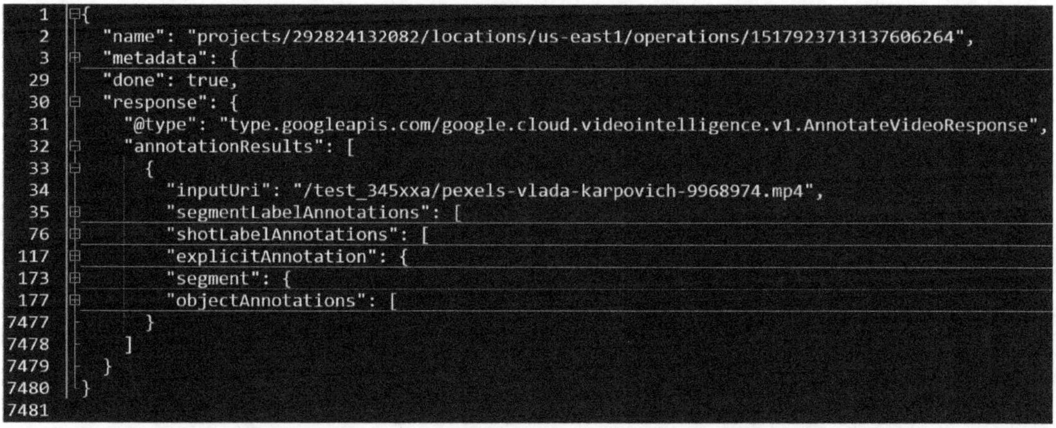

Figure 10.7 – Sections and line distribution of the Video Intelligence API response

This is an example of an **object annotation**. Notice how the coffee cup and its bounding box are detected in each video frame, and information about the segment is also provided:

```
"objectAnnotations": [
  {
    "entity": {
      "entityId": "/m/02p5f1q",
      "description": "coffee cup",
      "languageCode": "en-US"
    },
    "frames": [
      {
        "normalizedBoundingBox": {
          "left": 0.40747705,
          "top": 0.4910854,
          "right": 0.6224501,
          "bottom": 0.609916
```

```
          },
          "timeOffset": "0s"
        },
...
        {
          "normalizedBoundingBox": {
            "left": 0.435281,
            "top": 0.5009521,
            "right": 0.64407194,
            "bottom": 0.62310815
          },
          "timeOffset": "13.320s"
        }
      ],
      "segment": {
        "startTimeOffset": "0s",
        "endTimeOffset": "13.320s"
      },
      "confidence": 0.84742075
    },
```

Label annotations are returned by segment and by shot, as in this example:

```
    "segmentLabelAnnotations": [
      {
        "entity": {
          "entityId": "/m/01spzs",
          "description": "still life",
          "languageCode": "en-US"
        },
        "categoryEntities": [
          {
            "entityId": "/m/05qdh",
            "description": "painting",
            "languageCode": "en-US"
          }
        ],
```

```
      "segments": [
        {
          "segment": {
            "startTimeOffset": "0s",
            "endTimeOffset": "13.360s"
          },
          "confidence": 0.3399705
        }
      ]
    },
    {
      "entity": {
        "entityId": "/m/02wbm",
        "description": "food",
        "languageCode": "en-US"
      },
      "segments": [
        {
          "segment": {
            "startTimeOffset": "0s",
            "endTimeOffset": "13.360s"
          },
          "confidence": 0.7001203
        }
      ]
    }
```

In my personal experience, label detection usually works better in Cloud Vision, so I sometimes extract video frames as images and use Cloud Vision API to extract the labels, combining that information with that provided by Video Intelligence. These APIs are not alternatives since they can and should be combined to provide the best possible results.

Finally, explicit annotations are returned for each frame, including an offset and one of the six degrees of likelihood:

```
      "explicitAnnotation": {
        "frames": [
          {
```

```
                    "timeOffset": "0.554343s",
                    "pornographyLikelihood": "VERY_UNLIKELY"
                },

      . . .

                {
                    "timeOffset": "12.697213s",
                    "pornographyLikelihood": "VERY_UNLIKELY"
                }
            ]
        },
```

Using the Python client library is, once again, really easy. This is an excerpt that performs label detection on a local video file, synchronously waiting for the operation to end. Notice that the timeout should always be set to a time longer than the video duration itself or it will always fail:

```
from google.cloud import videointelligence_v1 as
videointelligence
video_client = videointelligence.
VideoIntelligenceServiceClient()
features = [videointelligence.Feature.LABEL_DETECTION]
with io.open(path, "rb") as movie:
    input_content = movie.read()
operation = video_client.annotate_video(
    request={
        "features": features,
        "input_content": input_content
    }
)
result = operation.result(timeout=90)
```

As you can see, we can also add video recognition capabilities to our code in a matter of minutes.

In general, the Cloud Video Intelligence pricing model is based on the number of minutes of video processed and the features requested. You can find more information about the Free Tier and costs by feature in the link included in the *Further reading* section.

And this concludes the list of AI-based services in this chapter. Now, let's work together on an exercise to see how easy it is to combine some of these services in our code.

Hands-on exercise

For this example, I used a dataset created by Phillip Keung, Yichao Lu, György Szarvas, and Noah A. Smith in their paper, *The Multilingual Amazon Reviews Corpus*, in the proceedings of the 2020 Conference on Empirical Methods in Natural Language Processing, 2020, `https://arxiv.org/abs/2010.02573`. The data files include thousands of reviews of Amazon products.

This example will process the reviews to find out which nouns are more frequent in positive reviews and which ones are present in negative ones. While this is quite a basic approach, it could be the first iteration of an application to provide information about which products our company should focus on and which ones we should consider dropping.

To create the input file for our exercise, I downloaded all six JSON files from the `/dev` directory of the dataset, available at `https://docs.opendata.aws/amazon-reviews-ml/readme.html`, which requires an AWS account to access it. I created a small Python script to choose 16 random reviews from each JSON file, meaning a total of 96 reviews in 6 languages, which are combined by the script into a single JSON.

I included the `consolidate_reviews.py` script in the repository for this chapter, so you can download the files yourself and create your own JSON file if you wish. You can even modify the number of reviews selected from each language to your liking. Just in case, I also included my own copy of `consolidated_reviews.json` for you to use. This is a random entry taken from it:

```
{
    "review_id": "en_0182997",
    "product_id": "product_en_0465827",
    "reviewer_id": "reviewer_en_0008053",
    "stars": "2",
    "review_body": "this is heavy. the tripod isn't that great
and actually is the heaviest part.", "review_title": "this is
heavy",
    "language": "en",
    "product_category": "wireless"
}
```

Our Python code for this chapter will perform the following steps:

1. Load `consolidated_reviews.json` from disk.
2. Iterate through all the reviews.

3. Translate the reviews to the target language if required (I configured it to be English by default, but you can change it if you prefer to use another one). The original `review_body` field will remain untouched in each entry, and a new property called `translated_body` will be created to store the translated text for each review. If the source and destination languages are the same, then the `review_body` value will just be copied to `translated_body`.

4. Run a **syntax analysis** and a **document sentiment analysis** for each review body.

5. Associate a global score to each noun that appears in one or more reviews and modify it depending on the sentiment scores in those reviews by adding the product of score and magnitude each time a noun is mentioned.

6. Finally, print the top 10 and the bottom 10 nouns ordered by the global score.

The full source code for this exercise is available in the repository for this chapter as `analyze_reviews.py`, together with my copy of `consolidated_reviews.json`. You can run it directly from Cloud Shell by copying `analyze_reviews.py`, `consolidated_reviews.json`, and `requirements.txt` to a directory and then issuing the following commands to install the required Python modules and enable the required APIs:

```
pip3 install -r requirements.txt
gcloud services enable translate.googleapis.com
gcloud services enable language.googleapis.com
```

Then, just wait for a couple of minutes for the APIs to become available and run the analysis:

```
python3 analyze_reviews.py
```

> **Note**
>
> This is just an example of how we can translate and analyze a set of text files. The scoring system is extremely primitive and was used to provide a fast result since a complete Natural Language analysis would make the code much longer and more complicated.

First, let's define our constants, including the path to the consolidated reviews JSON and the target language for our translations. Feel free to change their values if you prefer others:

```
REVIEW_LIST_FILE_NAME = "./consolidated_reviews.json"
TARGET_LANGUAGE = "en"
```

Then, we will load the consolidated JSON file from disk:

```
review_file = open(REVIEW_LIST_FILE_NAME, 'r')
review_list = json.load(review_file)
```

Next, we will then instantiate the translation and Natural Language clients outside of the loops to speed up execution and save memory by reusing them:

```
language_client = language.LanguageServiceClient()
translate_client = translate.Client()
```

Now, we will initialize a global dictionary to store the score associated with each name:

```
global_score = {}
```

Then, let's iterate through all the reviews and translate them if required:

```
source_language = review_data['language']
if source_language == TARGET_LANGUAGE:
  print("- Translation not required")
  review_data["translated_body"] = review_data["review_body"]
else:
  print("- Translating from {} to {}".format(source_language,
TARGET_LANGUAGE))
  translate_result = translate_client.translate(
    review_data['review_body'],
    target_language=TARGET_LANGUAGE
  )
  review_data["translated_body"] =
    translate_result["translatedText"]
print("- Review translated to: {}".
  format(review_data["translated_body"]))
```

Next, let's perform syntax and document sentiment analysis using text annotation. You might need to change the encoding type if you configured a different target language:

```
lang_document_type = language.Document.Type.PLAIN_TEXT
lang_encoding_type = language.EncodingType.UTF8
lang_document = {
  "content": review_data["translated_body"],
  "type": lang_document_type,
  "language": TARGET_LANGUAGE
}
# Anotate text for syntax and document sentiment analysis
lang_features = {
```

```
    "extract_syntax": True,
    "extract_document_sentiment": True
}
language_result = language_client.annotate_text(
    request={
        "document": lang_document,
        "encoding_type": lang_encoding_type,
        "features": lang_features
    }
)
```

Now that the language analysis is complete, we will create a list of nouns appearing in the reviews so we can quickly recognize them later:

```
noun_list = set()
for token_info in language_result.tokens:
    token_text = token_info.text.content.lower()
    token_type = token_info.part_of_speech.tag
    if token_type == 6: # If it's a NOUN
        if token_text not in noun_list:
            noun_list.add(token_text)
```

Then we will iterate through all the sentences, get their scores and magnitudes, find all nouns used in those sentences, and update their global scores:

```
# Iterate through all the sentences
for sentence_info in language_result.sentences:
    sentence_text = sentence_info.text.content
    magnitude = sentence_info.sentiment.magnitude
    score = sentence_info.sentiment.score
    sentence_score = magnitude * score
    # Split each sentence in words
    word_list = sentence_text.split()
    for word in word_list:
        word = word.lower()
        # Find nouns in the sentence and update global scores
        if word in noun_list:
            if word in global_score:
                global_score[word] =
```

```
                global_score[word] + sentence_score
        else:
                global_score[word] = sentence_score
```

Finally, let's use a pandas DataFrame to order the scores and print the top and bottom 10 nouns:

```
ordered_scores = pd.DataFrame(global_score.items(),
    columns=['word', 'score']).sort_values(by='score',
    ascending=False)
print("TOP 10 Nouns with the highest scores:")
print(ordered_scores.head(10))
print("\nBOTTOM 10 Nouns with the lowest scores:")
print(ordered_scores.tail(10))
```

While this is not a very precise analysis, it can help us understand which names appeared in sentences with the most extreme magnitudes and what product features users write more frequently about. The analysis could be much more accurate if we increased the number of reviews included if they referred to a single category or even a single product and if we cleaned up the list of nouns to remove false positives and other forms of noise. Again, this was just a small example to show you how quick and easy it is to translate and analyze texts using Google Cloud.

This is the result I got in my execution; we can see classics such as *quality, sound, or price* among the nouns with the highest scores, and *sticker, button, or color* among the names at the bottom of the list:

```
TOP 10 Nouns with the highest scores:
          word   score
89        quality   2.52
315       camera   1.66
154       love   1.45
155       story   1.45
115       sound   1.45
246   delivery   1.30
195       night   1.30
320       video   1.17
119       book   1.13
61        price   0.94

BOTTOM 10 Nouns with the lowest scores:
          word   score
275       sticker  -0.82
```

```
78        cover    -0.85
31        weeks    -0.89
66       button    -0.89
254      velvet    -0.98
132       waste    -1.13
135    purchase    -1.13
175       color    -1.13
260     download   -1.28
55          bit    -1.70
```

What's next

Now that we are done with our exercise, here are some tips and some food for thought:

- Translating and analyzing text or extracting labels from media can become expensive as the number of files increases, especially if we need to re-run our analysis repeatedly. How would you modify the code so that each sentence is only translated and analyzed once, regardless of the number of times that you run the script?

- I didn't use service accounts for this example on purpose; however, they are the best option for using this kind of API. How would you modify the code to run using a service account? Is there any cloud service that can make it to run this code as a service account without requiring tokens or keys?

- How would you validate words that can be nouns and verbs or nouns and adjectives at the same time (i.e., love, which appears on the preceding list)?

And now, it's time to wrap up.

Summary

In this chapter, we covered a few ML-powered services that Google provides through public APIs. First, we discussed the differences between unstructured and structured data, and then we covered how we can convert speech to text.

Then, we used Cloud Translation to get all our text files translated into the same language and went through the different language analyses that we can perform using Cloud Natural Language. In the next sections, we discussed how Cloud Vision and Cloud Video Intelligence can help us better understand the content of our images and videos.

Finally, we used a hands-on exercise to try a combination of Cloud Translate and Cloud Natural Language services to analyze a bunch of Amazon reviews.

This chapter is the last one dedicated to showing you how to extend your code using the best products and features provided by Google Cloud. We will use the final section of the book to connect the dots and show you how to build hybrid and multi-cloud applications that can run anywhere. The next chapter will focus on common design patterns for these kinds of solutions.

Further reading

To learn more about the topics that were covered in this chapter, take a look at the following resources:

- *AutoML: Train high-quality custom ML models with minimal effort and ML expertise*: `https://cloud.google.com/automl`
- *Speech-to-Text request construction*: `https://cloud.google.com/speech-to-text/docs/speech-to-text-requests`
- *Compare Cloud Translation Basic and Advanced editions*: `https://cloud.google.com/translate/docs/editions`
- *Cloud Translation supported languages*: `https://cloud.google.com/translate/docs/languages`
- *Cloud Natural Language pricing*: `https://cloud.google.com/natural-language/pricing`
- *Cloud Vision pricing*: `https://cloud.google.com/vision/pricing`
- *Cloud Video Intelligence pricing*: `https://cloud.google.com/video-intelligence/pricing`

Part 4:
Connecting the Dots –Building Hybrid Cloud Solutions That Can Run Anywhere

The last part of the book combines all the topics described in the previous chapters to provide you with a practical focus. First of all, we will discuss the different architecture patterns that can be used when designing an application or service for Google Cloud.

Then, we will use three real-world scenarios to showcase the process of modernizing or migrating them to Google Cloud and explain the focus areas and the challenges that we may encounter.

The final chapter includes a lot of useful information, including best practices, tips, and a list of the most common pitfalls so that you can improve your understanding and face migration and development projects with a much greater chance of success.

This part contains the following chapters:

- *Chapter 11, Architecture Patterns for Hybrid and Multi-Cloud Solutions*
- *Chapter 12, Practical Use Cases of Google Cloud in Real-world Scenarios*
- *Chapter 13, Migration Pitfalls, Best Practices, and Useful Tips*

11

Architecture Patterns for Hybrid and Multi-Cloud Solutions

After covering the different alternatives for running our code on Google Cloud and describing which of its services and APIs we can use to simplify our code, the last part of this book will be dedicated to best practices, implementation examples, and tips on how to build and run solutions while putting all these pieces together. All these topics will apply to either new solutions or legacy ones being modernized or migrated to Google Cloud.

In this chapter, we will first define what hybrid and multi-cloud solutions are and why they can be useful. Then, we will discuss the best practices for designing these kinds of architectures and, finally, we'll divide the architecture patterns into two distinct groups and go through each of these patterns.

We'll cover the following main topics in this chapter:

- Definitions
- Why hybrid and multi-cloud?
- Best practices for hybrid and multi-cloud architectures
- Types of architecture patterns
- Distributed architecture patterns
- Redundant architecture patterns

Let's get started!

Defining hybrid and multi-cloud solutions

We will dedicate the first section of this chapter to clarifying the differences between hybrid and multi-cloud solutions, two concepts that, despite being often used interchangeably, have different meanings.

Many organizations have their own private computing environment, where they host their services. An **on-premises solution** runs all its components in a private computing environment. No cloud environments or components are used.

When we want to start using a public cloud provider to run a new solution or to migrate an existing one, we can do it by choosing one of the following solution types:

- A **native cloud solution** runs all of its components on a single public cloud provider. No private environments are used.

- A **hybrid cloud solution** combines our private computing environment with a public cloud provider. Here, we will be deciding, for each component of the solution, if it will run in the private or the public environment.

- A **multi-cloud solution** uses two or more different public cloud providers to run applications and services, and may optionally include our private computing environment. Again, we will define the criteria to decide in which of the available environments each component will run:

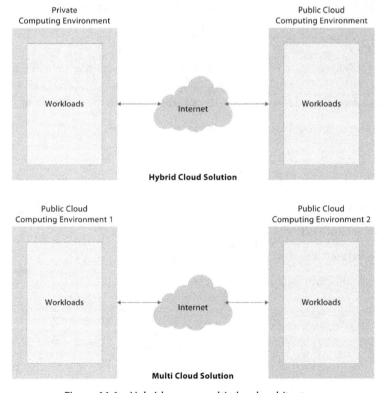

Figure 11.1 – Hybrid versus multi-cloud architecture

As you can see in the preceding diagram, a hybrid architecture interconnects a private computing environment with one or more public cloud providers, while a multi-cloud design interconnects two or more different public cloud providers.

Now that we understand the difference between the different solution types in terms of environments used, let's discuss which key topics we should keep in mind when we decide to use one of these architectures.

Why hybrid and multi-cloud?

You may be surprised to be reading about hybrid and multi-cloud applications at this point, after spending many chapters discussing how to build native cloud applications. The reason is quite simple: many organizations have many of their workloads still running on-premise and, even if you need to build a new architecture, you should consider hybrid and multi-cloud design patterns and compare the pros and cons of each approach before making a final decision.

While moving to the cloud is a priority for many organizations, many of them are still hosting their main services on-premises. In these situations, hybrid architectures can help us run at least part of any new solution on Google Cloud until all the core components our solution depends on have been migrated.

Hybrid approaches are also often used as a first stage when migrating to the cloud, moving first those non-critical services that have no dependencies. This approach reduces associated risks and can help the team performing the migration improve their skills and experience, something that will be useful when the time comes to migrate the most complex pieces of our corporate architecture.

Another possibility is to use Google Cloud to extend our on-premises computing environment during seasonal peaks or special occasions. Alternatively, we can set up a failover environment that will take control if our on-premises environment suffers an outage.

All the approaches mentioned have something in common: they can help us capitalize on all the investment in on-premises hardware, at least until it reaches its end of life, which is another reason to justify the use of these kinds of solutions.

A key part of the progress of migrating to the cloud is deciding which of our workloads will remain on-premises and which will run on Google Cloud. This is not an easy decision. The interests of the different stakeholders should be considered, and we should deliver an architecture guide that works for everyone. There are also many other factors to be taken into account, such as licensing and local regulations. A guide that considers all these conditioning factors should be put together as part of the migration, and it should not only include a migration plan but also let the organization know how to proceed from now on so that we get new workloads deployed in the proper location, thus avoiding extra migration work.

For example, if we decide that we will run all our frontends on Google Cloud, we will need to define a process to migrate existing frontends, but also another one to deploy new workloads directly on Google Cloud. Both processes will contribute to fulfilling our architectural vision.

Adding a new environment to our landscape will also bring added complexity to the table, so we will need to consider some areas where specific mitigation actions may be required. We'll discuss the most important ones in the next section.

Best practices for hybrid and multi-cloud architectures

In a hybrid or multi-cloud solution, we will have components running on different environments. The following is a list of some topics that we should keep in mind when we build a new solution or start modernizing a legacy one, in no special order. I have also added additional resources to the *Further reading* section, located at the end of this chapter:

- **Abstraction**: Being able to abstract the differences between environments will be key to establishing common tooling and processes across all these environments, sometimes at the expense of giving up on some unique features of a cloud provider so that we can achieve **workload portability**. This concept is the ability to move any workload from one environment to another and should be one of our objectives to pursue. Some examples of technical decisions that can help us in our abstraction process are using containers and Google Kubernetes Engine to standardize container deployment and management or using an API gateway to unify the different layers in our solution.

- **Processes**: Being able to use the same tools for deployment, management, and monitoring across the environment, as well as harmonizing our processes, will make our daily work simpler and speed up development, while also preventing undetected issues or environment incompatibilities from passing to production unnoticed. While this will involve additional efforts in the early stages of the process, it will be worthwhile later on. However, please ensure that the processes and tools don't become too complex; otherwise, the cure may be worse than the disease.

- **Synchronization and dependencies**: We should reduce or eliminate dependencies and synchronization processes across environments. These, particularly if they are running synchronously, may hurt the global performance of our solutions and are potential points of error in our architecture.

- **Data location and requirements**: While enabling workload portability should be one of our priorities, we shouldn't forget that the associated data has to "move" with the workload, sometimes just figuratively but others in a physical way. This means that we need to have a strategy in mind to make our data available to our workloads, regardless of the environment where they run. We can achieve this by either interconnecting our environments or by enabling data mirroring or data replication, each of which has its pros and cons that we should consider for each of our use cases before making a final choice. You can find more information about this topic in the *Further reading* section, located at the end of this chapter.

- **Connectivity**: Having multiple environments means that we will need to connect our private computing environment with one or more public cloud providers. For this to work properly, we should choose an appropriate connectivity model and a network topology that fits our use case that works for each of those environments. Performance, reliability, and availability are key metrics that we will need to consider before choosing the option that better fits our use case. For example, we can consider options such as Direct Interconnect or Direct Peering to help reduce our charges. We will match each architecture pattern, described later in this chapter, with a suggested network topology to use.

- **Security**: Security is always a key topic, and it will be more than ever when multiple environments are involved. There are some actions that we can take to help, such as encrypting data at transit, adding extra security layers using API gateways, or implementing a common identity space for all our environments, a topic that you can find more details about in the *Further reading* section. For example, if we use Google Cloud resources to run batch or non-interactive jobs, we should block external access to all infrastructure elements involved. Always remember to implement the **principle of least privilege** that we discussed earlier in this book to minimize potential security risks.

- **Cloud resources location**: Choosing our Google Cloud region wisely will help us reduce latency and maximize performance if we use one that is geographically close to our on-premises location. However, we will also need to study any additional requirements before making a final choice, depending on the use that we will make of the cloud. For example, if we will be using it as a failover environment, we may prefer it to be not so close.

- **Cloud cost model**: Knowing the details of our cloud provider's cost model can help us make better choices, too. For example, as we will discuss later in this chapter, some of the proposed architecture patterns for Google Cloud benefit from the fact that ingress traffic is free to offer a cost reduction.

Now that we have these best practices in mind, let's use the next section to divide architecture patterns into two groups before we go through the actual list.

Types of architecture patterns

There are two different types of architecture patterns, depending on our priorities and the reason why we have decided to use these architectures for a new solution or a newer version of an existing one:

- **Distributed architecture patterns** are those where we will decide, for each of our components, which environment where they will run. This means that we are prioritizing the performance, compatibility, and cost of each component over any other factors in our design.

- **Redundant architecture patterns** use multiple redundant deployments so that our components will be ready to run in more than one environment and may "*jump from one to another*" when it's required, focusing our efforts on increasing the capacity of our environments or maximizing the resiliency of our applications.

Which one should we choose? As you can imagine already, it will depend on the context, the requirements, and our objectives. As I already mentioned for other choices earlier in this book, in most situations, we shouldn't choose just one, but combine a few of them to get the most out of each of our environments, including Google Cloud, of course.

Next, let's enumerate and discuss the details of each of the different architectural patterns, starting with the distributed ones.

Distributed architecture patterns

There are four different patterns that we can use in our hybrid and multi-cloud distributed architectures. Each of them can be applied to different scenarios, depending on if we finally decided to use hybrid or multi-cloud, and depending on the current and future desired state of our application or service.

There will be a network topology associated with each of them that I will introduce, but you can find more details about each of them in the *Further reading* section, located at the end of this chapter.

Please remember that these are just patterns, so they may need additional customizations to suit each specific use case, but being aware of them will help us identify which of the available options can be used in a given scenario, and compare them to find the best one.

Now, let's go through each of the distributed architecture patterns.

Tiered Hybrid

The **Tiered Hybrid pattern** is an interesting option for organizations who want to migrate applications where frontend and backend components are decoupled, but it can also be a useful choice if we want to design an architecture for a new solution and our organization still has key components running on-premises:

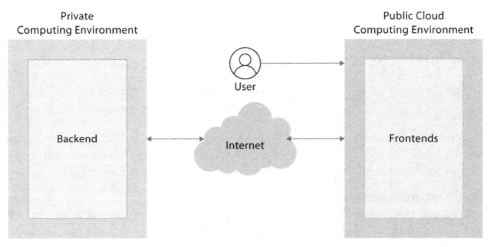

Figure 11.2 – The Tiered Hybrid design pattern

In its most usual implementation, in a tiered hybrid application, the frontend components run on Google Cloud, while the backend runs on-premises.

In migration scenarios, this can mean an enormous simplification because backends rarely depend on frontends, so they can be safely decoupled. Besides, frontends are stateless and can benefit a lot from the load balancing and autoscaling features of Google Cloud, so a public cloud provider is a better environment for them to run.

On the other hand, backends often contain data subject to local regulations, which may complicate their migration to the cloud, especially if the cloud provider is located in a different country, so moving the frontends first is usually a quick win. Finally, with this distributed setup, there should be much more ingress traffic from on-premises to Google Cloud (which is free) than egress, and this fact can help reduce our costs.

There is a less common reverse approach that's useful in migration scenarios, including legacy huge and monolithic frontends, where we may prefer to move the backend to the cloud first, and keep the frontend on-premises so that we can split it and migrate it later on.

In both cases, we should work on harmonizing the tools and processes to minimize the differences when deploying workloads to one side or the other.

In migrations, this approach can be used as a temporary step before deciding if the rest of the components of the solution should be moved to the cloud or not, all while reducing risks. Besides, the process of migrating or deploying the frontend can be used to learn more about the cloud provider and its service, and be better prepared to deal with the most complicated part of the migration later.

From a networking point of view, this architecture pattern requires stable connectivity between the frontend and backend, with low latency. One option could be a **gated egress topology**, where the visibility of components running on Google Cloud is limited to the API endpoints exposed from the private environment. A **meshed topology** could be another option, enabling all environments to connect with each other.

Next, let's use the next section to discuss how to use multiple cloud providers to host components of our applications.

Partitioned multi-cloud

In this architecture pattern, we have two or more public cloud environments from different providers and will decide the best destination for each component of our solution. This option maximizes our flexibility and can help us make the most out of each provider, while also avoiding potential vendor lock-in. It can also be useful if our organization has users in a country where some of our providers are not present, so we can use the others to fill that gap:

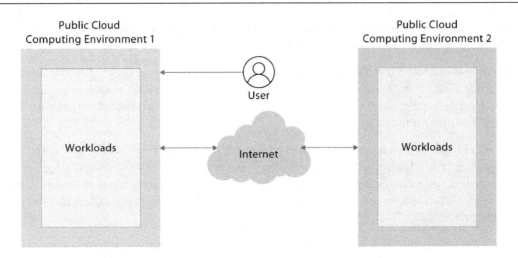

Figure 11.3 – Partitioned multi-cloud pattern

However, notice that for this pattern to be effective, we will need to maximize the abstraction of each platform and that this will require significant effort. For example, using containers and GKE can be a very interesting option that will enable the possibility of moving workloads between providers, helping us optimize our operations. Also, notice that some usual components in our architectures, such as load balancers, may not work if we plan to use them with multiple cloud providers and DNS will probably need some special setup, too. **Meshed or gated topologies**, which we described in the previous pattern, are the usual recommendation for these types of architectures too. In this case, gated ingress and egress are interchangeable, because we may want to expose the APIs from one environment to the other, or just interconnect them all.

We should carefully consider the additional complexities that this option may bring and decide if it is worth it for our specific use case. It is also a good practice to minimize the dependencies between components running on different providers, thus preventing an outage in one of the providers that may spread to the rest. Identity should also be unified if possible and security should be properly implemented and carefully monitored.

Partitioned multi-cloud is an interesting pattern, but some others are special. One of them, which we will describe in the next section, decides where our workloads should run depending on whether they generate or if they consume our data. How does it sound to you?

Analytics hybrid or multi-cloud

This architecture pattern can be useful if we want to deploy a new analytical solution or if we want to modernize our legacy setup. Since data extraction and data analysis pipelines are often separated from each other, this pattern suggests running the analytical workloads on Google Cloud, while data extraction is still done either on-premises or on the current cloud provider:

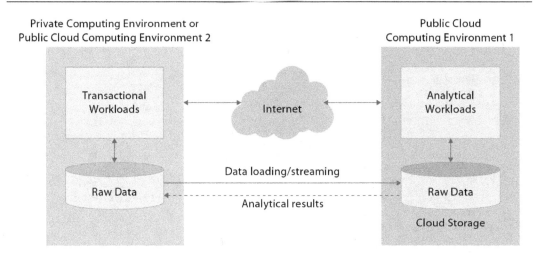

Figure 11.4 – Analytics hybrid or multi-cloud pattern

For **analytics hybrid or multi-cloud** to work, extracted data should be later copied to Google Cloud (and as ingress traffic, that would have no cost for us) where it will be analyzed. Depending on how our process is defined, the results of the analysis may need to be copied back to our on-premises environment where the original data came from.

This suggested split is reasonable because the analysis process is usually the most resource-intensive one, so if we run it on Google Cloud, we can benefit from a nice portfolio of data analytics solutions and products, and the availability of more resources with autoscaling capabilities, among other interesting features.

A **handover topology** is recommended for this kind of architecture, where connectivity to Google Cloud is limited, except for Pub/Sub and/or Cloud Storage, which are used to send data from the private to the public environment, either using data slices or messages to transfer it. This data is later processed inside the public cloud environment.

We can also add **gated egress** features to our topology if the volume of the resulting data that needs to travel back is significant or if using APIs is needed to send it back.

The next one is the last distributed pattern that we will cover in this chapter. It's a very interesting case because we often assume that connectivity is a critical service that should be reliable, but this will often not be the case. We'll discuss how to handle this and similar situations in the next section.

Edge Hybrid

Using the **Edge Hybrid** pattern makes sense for those use cases where we are receiving data from multiple locations intermittently or just at specific times of the day. These scenarios are special because internet access is not on the list of critical dependencies and data synchronization may happen at any time and can even be suddenly interrupted:

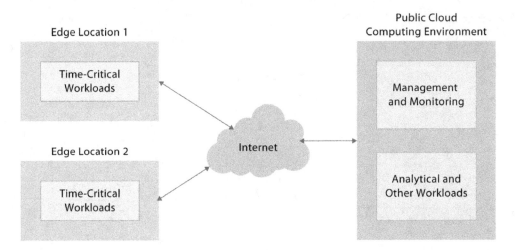

Figure 11.5 – The Edge hybrid pattern

We can also benefit from the free ingress with this architecture by using Google Cloud to receive and process all the data, and we should try to abstract the communication with all edge locations to make the process simpler. Using containers and Kubernetes, together with a set of harmonized processes, can help a lot in cases like this. We should also move as many of our workloads out from the edge locations and into Google Cloud, which will simplify our architecture and contribute to reducing the number of potential points of failure.

In this many-to-one architecture, data encryption and security will also be key factors to consider since sensitive data may be transferred between environments. **Gated ingress** would be a nice network topology for unidirectional setups, and **gated ingress and egress** would fit bidirectional scenarios. The decision will depend on which sides need to expose their endpoints and if we need to connect from the public environment to the private one.

This was all for the distributed patterns. Now, let's move on to the next section to discuss how redundant patterns can also be of help in many other situations.

Redundant architecture patterns

As we mentioned earlier, redundant architectures, as their name suggests, use copies of the same environment hosted by different providers in different locations. Different scenarios can benefit from redundancy, most of which can be mapped to the following architecture patterns.

Environment hybrid

Environment hybrid patterns run production workloads on-premises, while the rest of the environments, such as development, test, and staging, run on the public cloud provider:

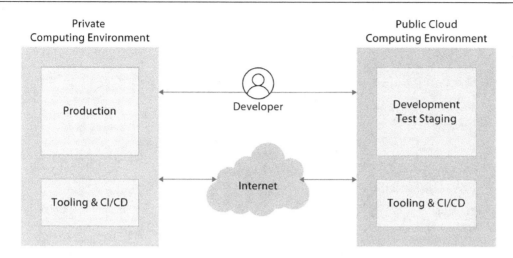

Figure 11.6 – The environmental hybrid pattern

This can be a nice way to speed up migrations because production is often the most challenging environment to migrate, but development, testing, and staging don't usually have so many requirements and limitations.

This pattern can also offer cost savings since non-critical environments can be shut down when they are not in use, and in that case, we will only be paying for the storage used by these environments.

For this pattern to work properly, all environments must be equivalent in terms of functionality, so abstraction is a priority here. GKE can be a useful option once again, combined with a set of tools and processes that are harmonized across all our environments. This means that a corresponding managed service should exist in Google Cloud for each product or service that we want to use or run on-premises.

A **mirrored network topology** is a nice choice for this pattern, where workloads running on the production environment are isolated from the rest of the environments. However, CI/CD and management processes are globally available and can be used in all environments, including a monitoring and logging system that is compatible with all our environments, wherever they are located, such as **Prometheus**.

Another classic design pattern that makes use of a public cloud provider sets up a latent copy of our production environments and turns it on when a disaster happens. We'll discuss it in the next section.

Business continuity hybrid or multi-cloud

Business continuity hybrid or multi-cloud presents another common way of integrating the cloud into our architectures, both at solution design time and when we are looking for ways to improve legacy applications:

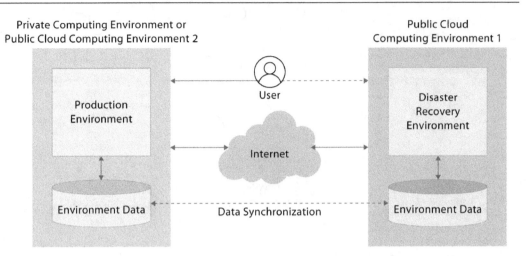

Figure 11.7 – The business continuity hybrid or multi-cloud pattern

In its hybrid option, we use a public cloud provider to set up a failover environment for our on-premises private environment. There is also a multi-cloud version, where our application runs on a first public-cloud provider and we use a second one to host our failover environment. In most cases, failover environments are built using virtual machines.

Using Google Cloud to set up a failover environment is a good choice because of its pay-per-use model and its presence in many different regions. We can also benefit from cost savings if we keep our failover environment on standby, only paying for the storage of our VMs. The ability to quickly resize our environment can also be of help during the execution of our Disaster Recovery measures.

Among the best practices recommended for this architecture pattern, the first one is to have an actual **Disaster Recovery plan** in place, carefully defined. Second, this plan should be periodically executed to properly test it, identify any existing issues and gaps, and improve it every year; otherwise, we may find unpleasant surprises when we need to use our failover environment in the middle of a crisis (and I have seen this happen to customers more than once, so you have been warned!).

Having a failover environment ready to run on Google Cloud whenever we need it will require us to back up or synchronize our data to the cloud, and we should ensure that this data is traveling encrypted, especially if it is sensitive. As part of this best practice, we should also decide if our data synchronization should be unidirectional or bidirectional. If this seems like an easy decision to you, you should read more about the **split-brain problem** in the *Further reading* section, located at the end of this chapter, to better understand potential risks and be better prepared to mitigate them. Besides this, our backup strategy should include additional destinations for copies of our data so that we don't only rely on our failover system when the main system is unavailable.

There are some other useful recommendations when implementing this pattern. First, there should be no dependencies between the main environment and the failover one. Containers and Kubernetes can be once again a nice choice to abstract the differences between environments, an abstraction that will be key to establishing harmonized processes, such as CI/CD, which should be replicated in each environment so that one is a mirror of the other and can easily replace it when it's down. Also, a common failover architecture will use a **handover network topology** for data backups and a **mirrored topology** for any other uses, both of which have already been described for other architecture patterns.

Load balancers are often used in this kind of architecture to redirect requests to one or another environment, depending on the circumstances, but we tend to forget that they can fail too, so having a plan B for them is a must, such as configuring our DNS with a fallback to the failover environment to ensure that connections are switched even when the load balancer fails.

Another pattern includes those cases where we use our cloud provider as an extension of our private environment to handle peaks. This pattern is known as cloud bursting.

Cloud bursting

Cloud bursting is one of the most common uses of public cloud providers to extend and improve existing applications but can also be used in new solutions to provide scaling capabilities to the often limited resources available on-premises:

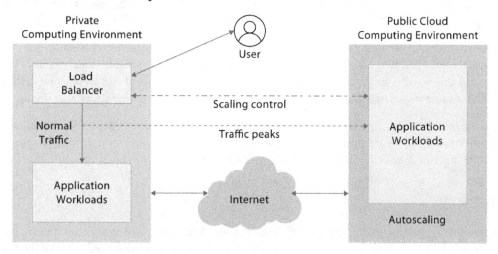

Figure 11.8 – The cloud bursting pattern

In summary, this pattern suggests running the baseline load of our solution in our private environment, and either switching to the cloud environment or complementing it with other instances running temporarily on Google Cloud only when extra capacity is needed.

For this pattern to be viable, we need a decent level of abstraction to enable **workload portability** so that we can move workloads to and from Google Cloud. A load balancer can be used to decide where each request has to be served from, but a normal one may not be able to read the currently available amount of cloud resources, and will also probably lack the ability to scale them up or down dynamically when required, even down to zero, which would be a desirable feature for dealing with peaks while minimizing costs. Being able to implement these features may complicate our implementation, so we'd better decide whether it's worth the extra effort.

Some organizations use this pattern as an intermediate solution before they complete their migration to the cloud, and run a cloud bursting architecture until the hardware running on-premises reaches its end of life. This is a nice way to make the most out of the Capex investments until they are no longer usable. Indeed, a good cloud bursting architecture should support a hybrid scenario, but also be ready to switch and just use components running on the public cloud at any time.

A **meshed topology**, which interconnects all environments, is a good networking choice to ensure that all workloads can access our resources, regardless of the environment where they are running. Using the right Google Cloud region to minimize latency and encrypting data at transit is also a good practice to succeed. And don't forget to supervise the amount of **egress traffic** that is required, because you will be charged for it!

As we have seen in other architecture patterns, we can achieve portability by using containers and GKE for lightweight workloads, and in this case, we can add VMs with a managed instance group for the most intensive ones.

Additional cost savings can be achieved in multiple ways. First of all, we will no longer need on-premises extra capacity to be reserved for peaks, since we can use Google Cloud to provide extra resources that scale up and down with our demand. Also, we can use preemptible VMs to run all our non-critical jobs, which can be restarted and take less than 24 hours to complete, with important cost savings.

This was the last architecture pattern for hybrid and multi-cloud architectures. I hope this list improved your knowledge and you are now able to consider different options when you need to design an application that runs totally or partially on Google Cloud. Next, it's time to wrap up.

Summary

We started this chapter by explaining the difference between the concepts of hybrid and multi-cloud solutions and discussing why these architectures make sense. Then, we enumerated some of the best practices to use when designing these kinds of solutions.

Next, we divided hybrid and multi-cloud architecture patterns into two different categories: distributed and redundant. Then, we discussed each of the different patterns that Google Cloud suggests to consider, explaining when they make sense and what we need to take into account if we decide to use them, including the recommended network topology.

In the next chapter, we will put all these concepts and options into practice and take three sample applications to the next level by containerizing them and making them hybrid and multi-cloud friendly.

Further reading

To learn more about the topics that were covered in this chapter, take a look at the following resources:

- *Understanding the difference between Mirroring and Replication*: https://wisdomplexus. com/blogs/difference-between-mirroring-and-replication/

- *Limits to workload portability*: https://cloud.google.com/architecture/ hybrid-and-multi-cloud-patterns-and-practices#limits-to-workload- portability

- *Authenticating workforce users in a hybrid environment*: https://cloud.google. com/architecture/authenticating-corporate-users-in-a-hybrid- environment

- *Eliminating MySQL Split-Brain in Multi-Cloud Databases*: https://severalnines. com/blog/eliminating-mysql-split-brain-multi-cloud-databases/

- *What is Prometheus?*: https://prometheus.io/docs/introduction/overview/

- *Hybrid and multi-cloud patterns and practices*: https://cloud.google.com/ architecture/hybrid-and-multi-cloud-patterns-and-practices

- *Hybrid and multi-cloud architecture patterns*: https://cloud.google.com/ architecture/hybrid-and-multi-cloud-architecture-patterns

- *Best practices for hybrid and multi-cloud networking topologies*: https://cloud.google. com/architecture/hybrid-and-multi-cloud-network-topologies#best- practices-topologies

## 12

Practical Use Cases of Google Cloud in Real-World Scenarios

We are getting to the end of the book, and it's the perfect moment to combine everything we've learned to implement solutions for three very different scenarios by analyzing their challenges and discussing how we can approach the design of the corresponding solutions, aimed at either replacing them or extending them, using one or more of the architecture patterns discussed in the previous chapter.

The process that we will follow is valid for building a new solution, but also for migrating a legacy application or modernizing an existing one. In all these cases, we can benefit from the exercise of comparing how developers traditionally used to face challenges, and how we can use the strengths of Google Cloud to create a better alternative, in terms of features, performance, security, and cost. I will also include a section at the end of each example with potential multi-cloud alternatives that could also be a nice fit in many cases.

Also, for your reference, and since there are hundreds of architecture patterns and guides that can help us solve specific problems and needs, I added a link to the Google Cloud solutions guide in the *Further reading* section, located at the end of the chapter. I'm sure you will find this resource useful sooner or later.

We'll cover the following main topics in this chapter:

- Invoice management – deconstructing the monolith
- Highway toll system – centralizing and automating a distributed scenario
- Fashion Victims – using the cloud as an extension of our business

Invoice management – deconstructing the monolith

Our first scenario is a legacy invoice management application running on a Windows server. This is a huge monolithic application with multiple features that eats up more and more RAM and has become a challenge to maintain due to its frequent crashes.

As part of the corporate IT modernization plan, we are assigned the task of modernizing the application. We have access to the source code and have a team of developers to work with us who have been upskilled to develop on Google Cloud with some online training and a copy of this book :) We are now ready to take this application to the next level, so let's do it!

Specifications

After some interviews with the developers and the team supporting the application, and a careful look at the source code, we have identified the following key features and details of our invoice management software:

- It runs on a Windows server using a single binary.

- It uses a single external SQL Server database to store application data.

- It has a graphical UI that is used by members of the accounting team around the world using a **Remote Desktop Protocol (RDP)** client to connect to the main server, where the binary runs. Only one user can use the UI at a given moment.

- It allows the creation of new invoices from the UI, as well as modifying them and marking them for deletion, but no invoice can be physically deleted for audit reasons.

- Signed PDFs can be generated for any invoice.

- It has an integrated payment system that uses an external API to send money via wire transfers.

- Since the company has providers from different countries, the solution supports multiple currencies, obtaining updated exchange information periodically from an external third-party provider, using its own API.

Analysis and opportunities

While we should get a lot more information before moving on, the previous list gives us some very interesting hints about how we can proceed.

This is a diagram of the original architecture, which uses an active-passive Windows cluster to host the application and the database in different nodes:

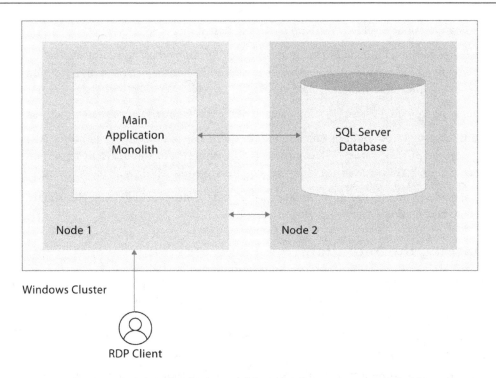

Figure 12.1 – Original architecture of the invoice management application

Running a single-process application on a unique server, even if it's configured in high availability using a cluster, has quite a few risks. First, an error or an exception in one of the threads may affect the whole application in cases such as—for example—a memory leak.

Also, the user experience can be affected by the latency of the RDP connection, which may be particularly annoying when we are connecting from a location geographically far away from the application server.

Deploying a new version of the binary with added features can also be a problem because it will not be possible to do so without an associated downtime, and since we have users all around the world, it will be complicated to find a proper time that does not interfere with daily work.

Besides, the current use of the application is limited to a single concurrent user, since RDP connects to the session running the binary, and this is a restriction that we should eliminate, together with the dependency on RDP.

Some improvements that we could present for this specific case if we rebuilt the application to run on Google Cloud are listed here:

- UI modernization, with multiple users able to connect and use the application at the same time
- Infrastructure that autoscales to adapt to the number of connected users, scaling down to zero when it's not in use

- Deployment of new versions with no associated downtime

- Replicated local UI and database instances in any location where the company has an office (if Google Cloud has a presence; otherwise, we would choose the closest one) for minimum latency and the highest possible performance

- Feature parity with the legacy version—signed PDFs, wire transfers, and currency exchange rates

- Additional features could be enabled thanks to the use of Google Cloud public APIs and services—**Optical Character Recognition (OCR)** scanning to automatically import invoices or an external API made available for our suppliers to add their own invoices remotely

- Cost savings due to the pay-per-use model, autoscaling, and getting rid of hardware maintenance tasks

Approaching the migration

The first question that we should ask our managers has to do with time. How long can the application run in its current state and location? Getting an answer is important to help us understand whether we need to perform a fast migration or whether we can take our time to define a careful plan to be executed step by step. Hardware end-of-life information and other corporate information should be taken into account at this stage.

Another important question is this: Can the data stored in the database—particularly financial and personal information—be moved to the cloud or not, depending on local regulations and the location that we are choosing in Google Cloud? If the answer is no, our only cloud-friendly option would be to use a hybrid approach, where the database would remain on-premises and data would travel encrypted to the frontend, benefiting from the no-cost ingress. I will assume that this is not the case and that we have the green light to move all components to Google Cloud.

If we are almost out of time to complete the migration, we should either use a lift-and-shift bare-metal approach for the hardware or virtualize the servers. Before going virtual, we should check whether the current amount of memory used and the number of CPUs required to run the applications are available on Google Cloud, and run some tests to confirm that there is no performance degradation, particularly in the database.

Lift and shift could also be a quick first step to follow to earn some time, and once the availability of the application is guaranteed, we can start working on our future vision for the application in less stressful circumstances.

If time is limited but there's still a bit left, we could choose to build a **minimum viable product (MVP)** on Google Cloud, with enough features to keep our users working, and then work on additional features on a periodical basis. This approach can be risky and will require a well-crafted testing and validation plan for the MVP before we can switch from the legacy application. A CI/CD system and related tooling should also be defined as part of the design so that we can deploy newer versions quickly and without requiring any downtime.

We may choose the MVP approach even if we have plenty of time ahead because it can help us validate that we are following the right path in terms of features and performance before it is too late to accommodate any fundamental changes in its architecture.

Working in sprints can be an interesting approach to prioritize both new features and bug fixes and work with the team on what matters most for both our users and our business at any given time.

Designing the new architecture

Once we have a plan to follow, it's time to start drafting the architecture of the new application, taking into account all features and requirements.

In this example, we have access to the source code of the original application, and one of the most important decisions to take is how much of this source code we are going to reuse for Google Cloud. We may choose to reuse as much as possible or otherwise refactor the whole code base. As we discussed at the beginning of the book, when writing code to be run on the cloud, we should keep security in mind from the very beginning, so if writing code from scratch does not make a big difference in terms of complexity and delivery time, it will be a better option for monolithic applications.

Our example has some clearly separated functionalities bundled into the monolithic binary, so we can start by splitting the monolith into smaller services and microservices, and we can then customize each of them to benefit from the special features that Google Cloud provides.

First, the frontend should be decoupled from the backend and should become multi-user friendly. Using a web browser to connect to the application could be an easy option that would work for all users.

Then, there should be a dedicated service for invoice management, including creation, modification, and marking for deletion. This module could be extended with another one to handle OCR for imported invoices using Document AI.

Another dedicated service should be deployed to handle payments using wire transfers to a specific account name and number. Note that this module should have extra security in place since an unwanted intrusion could have catastrophic implications. Of course, a project-wide identity system would be a good feature to enable a permission-based system to limit what each user can do and which modules or services they can use.

Separate services could also be implemented to periodically refresh and offer currency exchange rates and to generate signed PDFs. The former would need internet access, which should be limited to connections from that specific service to the host exposing the API.

Regarding data storage, we could use a Cloud SQL instance for all transactional data, but we could also complement it with Cloud Storage to store invoices and generated PDF files. Using Cloud Storage offers a few benefits: we can use multi-region buckets, which is compatible with our intention of getting geographically closer to our customers, and we can also build a token system to offer direct downloads of generated PDF files from the bucket using authenticated URLs.

Combining all these ideas, this would be a high-level draft of our new architecture:

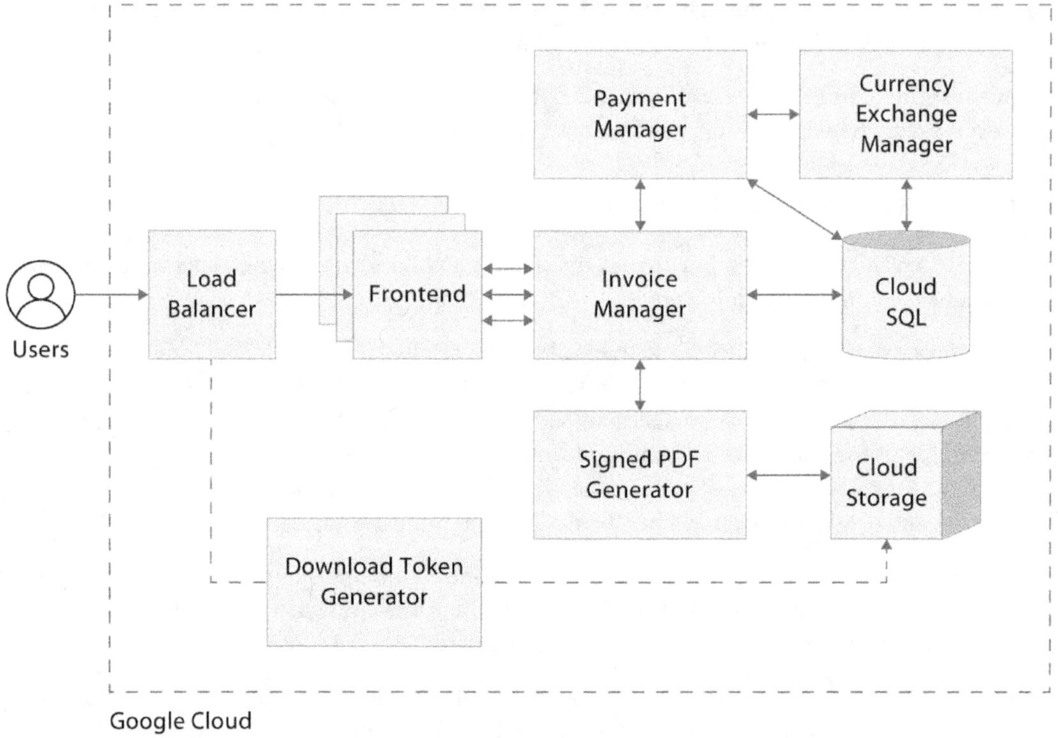

Figure 12.2 – Draft architecture for invoice management in Google Cloud

Looking at *Figure 12.2*, we can identify a few components that would benefit from being elastic. The frontend is an obvious choice so that we can have enough resources to serve multiple users, but also scale to one or zero when there is none. Since each connection to the frontend will interact at least with the invoice manager, this component should also be elastic to grow or shrink as the frontend does.

The currency exchange manager will provide a file for each request and should periodically update the data from the main source, so it shouldn't be a problem unless we have many users. A service such as Cloud Scheduler could be used to schedule periodic updates, and a Cloud Functions or a Google App Engine instance could be used to run its main functionality. Depending on how often we update currency rates, we could benefit from the fixed cost of App Engine and use it to run our scheduled jobs.

The signed PDF generator is in a comparable situation; it shouldn't be usual that multiple requests happen at the same time, but it will depend on the number of users. Using a Cloud Function could also work in this case, given that we are expecting a short execution time and a limited number of executions.

And related to the number of users too, we should decide whether this architecture would make sense in a unique location or whether it should be distributed across the regions where our users have a presence. If we decide to use a unique location, we should carefully decide which would be the best location to serve customers connecting from multiple countries.

If we otherwise decide that the architecture should be replicated in multiple locations, we should determine the number of locations and where these will be and decide whether we will need a central database, or whether an independent one per region will work. Note that using a centralized database would require periodical synchronization of available data in each regional instance, and user and permission management may also be affected by this decision.

Added features are often provided at the cost of increasing the overall complexity of the architecture, and this is a particularly good example of one that should have its pros and cons carefully considered before deciding whether it's worth the extra development time and complexity that it will add.

Hybrid and multi-cloud options

This scenario can also be implemented using a multi-cloud approach. You may choose this method if you have any existing components already running on a provider that cannot be moved, or if you just want to use various products that you like, each of them available in a different cloud provider. You may also have an agreement in place or pre-paid services that will take a while to expire. All these cases will be compatible with this scenario.

If we use **Google Kubernetes Engine** (**GKE**) for our services, we can use Anthos to deploy them to any of the compatible providers. This would simplify our architecture by centralizing management and unifying processes, which could be a great help given the usual complexity of multi-cloud architectures.

APIfying our services could help too—that is, using an API to provide a service instead of making our code talk directly to the database. For example, we could provide a method to list the documents for a user by just passing the user ID and returning a list using a predefined structure. The main benefit is that we don't need to expose the database port but just an HTTPS-secured port that offers limited functionality, which could help mitigate security risks if the code used were written following some of the security principles that we discussed earlier in the book.

With this approach in mind, we may decide to run the frontend on one public cloud provider and the backend on another. If we just need to expose our backend endpoints and invoke them from the frontend, setting up security and identity will be much simpler, and we can just block connections to any other components to improve security globally. On the other hand, the added complexity of distributing components from the same part of the application across different cloud providers will probably void any potential technical advantage, so I'd recommend avoiding it.

I hope you found this analysis interesting. Now, let's move on to the second example—a quite different scenario where we will also have a lot of room for improvement.

Highway toll system – centralizing and automating a distributed scenario

In our second example, our company signs a contract with the local government to take care of the maintenance of all the toll highways in the country for 10 years in exchange for all tolls collected from vehicles using them.

Specifications

There are 537 toll booths in the whole highway network, distributed at different exits where there are usually 2 or more booths in each. Our manager needs to know how much money was collected each day and be able to see a dashboard with a historical evolution of the total amount collected daily.

The company is currently using the following process:

1. Each toll booth has a small computer with an Ethernet connection and limited storage, where each transaction can be logged. There is also a camera for license plates and a weighing scale. The booth also has a private screen that displays the total amount of money collected and is secured with a cover and a lock. Software running on the booths can be customized since we have access to its source code.

2. When a car enters a toll highway, the driver either gets a paper ticket or uses an electronic device to open the entrance gates. Electronic devices are provided to subscribers, who pay their consolidated tolls at the end of the month. At exit time, drivers either use their device again to open the exit gate or insert the paper ticket they got when they entered and pay the due amount in cash or by using a credit card. This amount depends on the vehicle's type and weight.

3. Each highway exit has a manager who visits each booth at 1 A.M., opens the cover with a key, and takes note of the amount collected in each booth. Then, they go back to the office and sum the amount for all booths using a spreadsheet and attach it to an email, which is sent to the global manager using a **high-speed packet access** (**HSPA**) mobile connection available in every office, which is enabled from 1 A.M. to 3 A.M.

4. The global manager arrives at the office at 7 A.M., opens all 250+ spreadsheets, pastes them into a global spreadsheet where the global collected amount is calculated, and then deletes the rows for the least recent day. The global sheet is then sent to the vice president around 10 A.M.—this includes a line graph to see daily historical information for the last 365 days.

5. This vice president happens to also be the IT manager and wants you to modernize this system because updates were not received on time on 50+ days this year, due to different issues in booths or during the consolidation process performed by the global manager.

6. Installing additional software either in the booths or in any of the offices located at the exits is an option since they are all running remote connection software with file-transferring capabilities.

Analysis and opportunities

The combination of multiple, distributed locations and a lot of manual work makes this a perfect opportunity to use Google Cloud to centralize and automate all the described processes.

There are some serious issues in these scenarios. First of all, writing down collected amounts manually may lead to human errors, which could hamper the correct visibility of earnings. The company is also lacking insight into other events—for example, data from sensors such as weighing scales, payment method insights, payment system error rates, frequent customers, and other very valuable information that is currently being wasted.

Having the raw logs for all booths available in a central location would open a new world of opportunities for this company. New data could be identified to be logged in booths after an in-depth analysis of the whole process, and the company could start using all available data for strategic and marketing purposes.

This is a list of some of the improvements that we could offer for this specific scenario:

- Fully automated raw data ingesting process, with no human intervention, leading to fewer errors
- Failure-resistant design, where connectivity issues will not involve data loss, but just a delay in its ingestion
- Data processing will be performed on Google Cloud
- Self-service real-time dashboards that are available at any time for specific named user accounts, removing the limitation on the historical window
- Raw data will be stored in a central database, where it will be available for additional analysis and for fueling new projects, such as a traffic prediction model or a customer segmentation system to improve marketing communications
- An improved security model, not only for booths and offices but also for the centralized data lake

Designing the new architecture

If I had to choose a design pattern to implement a solution for this scenario, I would combine two of the patterns that we covered in the previous chapter to implement this solution: **Edge Hybrid** and **Analytics Hybrid**.

> **Note**
>
> As happens in real life, for each of the scenarios in this chapter, there will often be more than one possible solution. You will have your own ideas and may not agree with the choice I made, and that is totally fine, but I still wanted to show you what I would do for each so that you can use it to learn how to approach similar situations. In a real project, you will have to talk to different stakeholders and find the best solution for all of them, which will probably make decisions more complicated.

I would choose Edge Hybrid because of how booths and offices are distributed, and the fact that a similar sequence of actions is followed at each highway exit, so we can consider them as edge locations. Information is then consolidated in the corporate HQ at a specific time of the day, meaning that connectivity is not a critical service for these locations, which is compatible with this design pattern.

On the other hand, offices will be just pushing data to the cloud, and all the processing and insight extraction will be done on Google Cloud, which clearly resembles an Analytics Hybrid pattern.

This is, indeed, a very good example of how patterns are not mutually exclusive but often work very well in combination, and these combinations will make more sense when we are dealing with complex scenarios.

Once we have identified the design patterns involved, let's proceed to design the new architecture, taking into account every possible component that we could improve in the process.

First of all, let's imagine what happens when a car arrives at a toll booth located at one of the entrances to a toll road. If the driver is a subscriber, an electronic device is used to open the gate and an `entrance` event is logged containing UTC date and time, booth number, vehicle weight in pounds, and device ID. Otherwise, the driver picks up a ticket, and an `entrance` event is generated that includes UTC date and time, booth number, vehicle weight in pounds, and a ticket number. In both cases, a picture of the license plate is also attached to the event.

A couple of sample events, in a semicolon-separated CSV format, might look like this:

```
2023-05-12;09:05:12;14;2800;T6512;T6512-20230512090512.JPG
2023-05-12;21:53:33;98;4400;D0343;D0343-20230512215333.JPG
```

The first line belongs to a compact car entering the highway at booth 14 at around 9 A.M. and getting a ticket (notice how the ticket number begins with `T`). The second one is a large car entering the highway at booth 98 a bit earlier than 10 P.M. Both events also include the name of the file containing a photo of the license plate.

When a car exits the toll road, it will also arrive at an exit booth. At that point, the car could have a monthly pass, whereby the gates will automatically open. This could be logged as a payment event of the `payment` type with a `subscription` method and the corresponding amount to be paid at the end of the month, also logging the ID of the electronic toll device located in the car and the name of the file containing the license plate picture. The event will also contain the booth number and the exit number, together with the date and time. All this information can be used later for different analyses.

For non-subscribers, payments can be done by credit card or in cash, so we can log a `payment` event type, with the method being either `cash` or `card` and the corresponding amount. For credit card payments, we must log the card number, since charges will be performed the next morning.

Exit events for the cars mentioned in the previous example would look like this:

```
2023-05-12;10:15;49;2800;T006512;PAYMENT;CARD;ERROR;4,85
2023-05-12;10:18;08;2800;T006512;PAYMENT;CASH;OK;4,85
2023-05-12;23:43;82;4400;D000343;PAYMENT;SUBSCRIPTION;OK;8,36
```

The first car tried to pay $4.85 at 10:15 A.M. at booth 49, but something happened when trying to pay by card, so the driver finally decided to pay by cash at 10:18. The driver of the second car used a device to exit on booth 82, generating a pending payment of $8.36 to be added to the corresponding monthly bill. Notice that these events are missing the credit card number in the first event and the filename of the license plate photo in all three of them, which I redacted to make each event fit in a single line for easier reading, but this information would be included, too.

Now that we understand what events look like, let's design a system where each payment, together with its corresponding data, originally stored in the local booth storage area, is an event that must arrive at Google Cloud. New events are queued locally, and we use a **First-In, First-Out (FIFO)** process to send them to the central computer in the office. Whenever there is connectivity between each booth and the central computer, events are transmitted and their reception is confirmed by the destination—one by one—before they are deleted from the local storage of the booth.

This synchronization process could happen constantly, or just be triggered at a specific time of the day, if we can guarantee to have enough local storage for data of a few days, preventing connectivity issues from causing a loss of data. For instance, we could start the synchronization every night at 2 A.M., expecting all transfers to be completed at 3 A.M.

Another good practice is not to delete any information but to mark it for deletion, and use a separate process to delete stale data when more room is required. This will increase our chances of recovering data if any issue is detected.

For this process to work properly, we should complement our architecture with a monitoring system that will check for connectivity errors and alert the office manager so that these can be fixed as soon as possible.

Once we have new events in the central computer in each office, we need to transfer them to Google Cloud. An interesting option to do this would be to use **Pub/Sub**. We can create a topic for booth events and start queueing our events. My recommendation for scenarios such as this is to implement our own code for acknowledging each message received so that it is deleted from the source only when we know it has been correctly ingested. Again, we can just mark a message for deletion and use a separate process to check when we are low on storage space, and only then proceed to the physical deletion of old event records.

Now, let's look at the event transmission process from Google Cloud. We receive a new event using Pub/Sub, which may trigger a Cloud Function to receive it if the amount of expected messages is limited, or Dataflow if streaming is required due to a high volume being expected. The Cloud Function or Dataflow could insert a row in a sharded table in BigQuery, where we have a separate table containing all the events for each day. The photos of the license plates would be stored in a Cloud Storage bucket and their filename would be included as one of the fields for the inserted row. At this point, we could trigger another Cloud Function to use the OCR feature of **Vision AI**, or even our own trained **computer vision/machine learning** (**CV/ML**) model, to read the license plate number and update the corresponding event record.

From the point of view of networking and security, there are a few points that we should keep in mind:

- Booths must have physical and perimetral security as we must guarantee the integrity of payment data since credit cards will be charged the next day. Physical and remote attacks must be prevented by isolating the network. The connection at the office may be the weakest link and should be properly protected from intruders.

- Booths should connect to the computer in the office just for two purposes: in one direction to send events to the computer located in the office, and in the other to provide a response to monitoring and failure-detection test requests. Any other connections should be dropped using a software firewall or a similar security solution, where an exception would be temporarily added during software updates. Failure monitoring could be performed using a self-test API or a URL exposed in each booth, in order to limit the number of connections required to complete the process.

- The computer in the office needs to be connected to the internet, but only to send Pub/Sub events during a specific time range, so we can use another firewall to enforce this limitation, thus preventing malware infections or any kind of threats.

All these requirements are compatible with handover and gated network topologies, which limit connectivity to just a few exposed API endpoints and a number of specific Google Cloud services, so they would be good options to implement in this case.

With all the mentioned topics in mind, this could be a diagram for a proposed architecture:

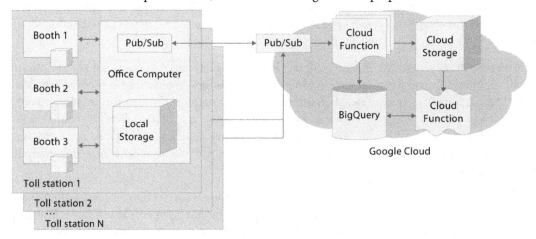

Figure 12.3 – Sample architecture diagram for the highway toll system

Making the most of our data

Once we have license plate and credit card numbers in our tables, we can use them to identify recurring users, an analysis that can be interesting for statistical purposes. We can also offer users a 10% monthly repayment if they register their license plates and credit cards on our website, which would allow us to match cars to their owners and open a new world of marketing opportunities.

We can also use Looker Studio to visualize our data and provide dashboards and ML to extract insights and perform predictions. Here is a list of ideas I could think of, in no particular order. I'm sure you've already thought of many others to add:

- Sort exits by money collected, by the number of cars, and by the division of both, in order to understand which ones are bringing in more money. We could use this information to prioritize booth extension or modernization processes.

- Study traffic patterns and volumes throughout the day to understand whether there are capacity issues and whether these affect just specific exits or all of the road network. This could help prioritize where additional efforts could be required, such as express lanes or more booths, or just to increase the number of subscribers using electronic devices to speed up toll collection. Capacity and route information could also be used to find which would be the best locations for services such as oil stations, rest areas, hotels, restaurants, and so on.

- Cluster vehicles using lifetime value based on recent purchases to identify high-value customers. These customers could be offered more aggressive subscription offers when they register since they use toll roads regularly. Drivers also need gas, new tires, and—eventually—to renew their cars, so we could also partner with oil stations, garages, and car dealers to make additional money while providing useful additional services to our customers.

- Obtain information about which car brands and models appear more frequently, using the information provided by users when they register their vehicles. Understanding vehicle sizes could help the company build toll booths that work for everyone, and this aggregated information could also be useful to car manufacturers, or even for the government.

Hybrid and multi-cloud options

Imagine that all toll booths are connected to one cloud provider and you need to run your analysis on another provider. Or, it could be that the data lake is hosted on one provider, but the analysis and the dashboards should be performed on another because you will activate insights in a compatible digital marketing platform. These are real-world cases in which using a hybrid or multi-cloud approach could make sense, especially for the second scenario.

In cases such as this, Google Cloud is ready to be used as the platform to run your analytical workflows, even if your source data is stored on **Amazon Simple Storage Service** (**Amazon S3**) or Azure Blob Storage using BigLake tables. BigQuery Omni is a product that can help us perform this analysis by deploying a BigQuery compute cluster in each additional environment, returning the results to Google

Cloud using a secure connection so that we can access the results of different analyses performed in **Amazon Web Services** (**AWS**), Azure, and Google Cloud and combine their results to provide visualizations and extract insights from our data.

This can be a very interesting option for big companies that are using different cloud providers and need to break data silos or run processes that require access to data spread across different providers. Using a product such as BigQuery Omni can simplify this process a lot because of the multiple complexities that integrating these processes may entail. Otherwise, an Analytics Hybrid design should be able to work for any other not-so-complex scenarios.

I hope this example has helped you understand how complex scenarios can be improved by combining different patterns and topologies in Google Cloud to build an architecture that is faster, less prone to errors, and more secure, and that can provide many additional features and insights based on data analysis to help businesses grow.

Next, let's see an example of one of the most common situations that can make use of Google Cloud to mitigate specific issues while reducing costs.

Fashion Victims – using the cloud as an extension of our business

I included this third scenario because it represents a common first contact of many companies with the cloud and, as we will discuss in the next and last chapter of the book, some of the decisions that we will be taking during the earliest stages of establishing our presence in Google Cloud will have an important impact on its future and may contribute to its success or otherwise totally ruin it.

This example has to do with a well-known *brick-and-mortar* brand. Fashion Victims has been welcoming its very exclusive customers for decades in a very representative three-floor concept store in the heart of Paris, France. Last year, the founder died, and his two daughters, who are now in control of the company, want to take the brand to the next level, opening their fashion catalog to everyone with more affordable offers for all tastes and budgets, and extending their presence to more than 10 new international markets by opening factories in each country that will produce the clothes locally.

Beyond the business challenges that such a change entails, which are way beyond the scope of this book, the priority of the new owners is to boost online sales. For this purpose, a new website and an app have been created, offering the full catalog in each country where there is a factory, plus special flash sales with amazing discounts and free shipping on all local orders.

A French hosting provider is used to power these services. Once the new website was live, the news went viral, and everyone wanted to become Fashion Victims' customers… And this is when the IT nightmare began.

This is what the current architecture looks like:

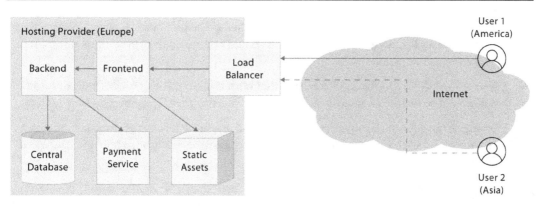

Figure 12.4 – Diagram of the current architecture for Fashion Victims

Now, let's take some time to study the current situation and identify potential opportunities for improvement.

Specifications

As we take a look at the analytics for the last 3 months, we can see a regular traffic pattern during normal days, but when a flash sale begins, the corresponding local site gets 10,000% hits, and servers start to fail due to their lack of resources to deal with the extra traffic. At that point, a custom error page is shown, asking customers to try again later, which makes both Fashion Victims and its customers frustrated—the former because of the loss of sales and the latter because they cannot buy their beloved garments.

Given that there are flash sales in different markets almost every week, we are urged to find an alternative to properly handle peaks without interrupting the service. Also, some customers from Asia and America are complaining about the overall performance of the website and the app, a problem that they are experiencing even when there are no active flash sales.

Buying additional hardware could be a first idea, but it would increase capital and operational expenses, and new equipment would be idling almost 65% of the time. We need an additional capacity that we can switch on and off at will, depending on the current level of traffic.

Luckily, Google Cloud can help us in this scenario, too. Let's see how.

Analysis and opportunities

This scenario is open to a lot of potential improvements. Let's enumerate some of them, as follows:

- The capacity issue is the main headache. We can use elastic cloud services to provide additional capacity, which could even scale down to zero when they are no longer in use. The most important part of this architecture will be the **load balancer** because we need to have accurate information about how much traffic we have and how many cloud resources we are using at any given moment so that our system can decide whether the traffic can be served by just using the on-premises infrastructure, or whether it should be handed over to additional resources deployed on Google Cloud.

- Serving users from Google Cloud means that both dynamic and static content should be synchronized so that all users see the latest content, wherever they are connecting from (and to).

- We can use the opportunity to minimize latency and maximize throughput by using the closest cloud region for each user, for the sake of additional content synchronization or caching. We will discuss this topic shortly.

- Last, but not least, we can reduce costs by moving toward a pay-per-use model and benefit from the competitive pricing of Google Cloud when compared with the costs of traditional hosting. It should not be a surprise that many customers take the next step and fully move to Google Cloud for hosting their properties after experiencing its benefits firsthand.

Designing the new architecture

If you remember the patterns that we discussed in the previous chapter, this may seem a perfect fit for a cloud-bursting architecture, where the cloud provider is used to absorb extra traffic during peak times. However, as I already mentioned earlier in this chapter, real-world scenarios are often not so simple, and this is one of those examples because we will need to understand better how data is used before we can decide whether we need to use additional design patterns or not.

For example, imagine that there is a central customer, inventory, and orders database hosted on-premises and subject to legal requirements that prevent personal data from being moved to a different country. This situation may lead to different potential scenarios.

In the first scenario, both the database and—optionally—part of the frontend would remain on-premises, and we would need to connect this environment to Google Cloud with decent network latency and performance—numbers that obviously would get worse as we connect from locations geographically far away. In this case, we may want to turn the database calls into API calls if they weren't already so that we could implement an additional gated topology to connect private and public environments, thus improving overall security. In this scenario, ingress traffic would be much more predominant, and we could benefit from the fact that it has no cost in Google Cloud.

A second option would be to move the central database to Google Cloud, which would basically mean switching ingress and egress traffic; this would increase costs because, in this case, egress traffic would be predominant. If we chose this option, it would probably make sense to move the frontend to the cloud too, and the advice to APIfy the database services would also apply.

A third option would be to build a Google Cloud self-contained architecture in each region, including the frontend and backend, and then periodically synchronize new orders from each regional database to the main one, which would be used for centralized reporting. For this synchronization, we could use a process similar to the one described for the toll collection example. A well-built elastic architecture including a small database in each region might not be so costly; it would make it easier to manage local inventories and would also help reduce dependencies across regions, which would cut down costs and also make it easier to open new locations as the company scales up.

For the sake of simplicity, I will choose an intermediate solution, aimed at controlling costs while keeping the central database on-premises. Self-contained sites will run on the cloud region closest to each factory, and users will be forwarded to the closest site by default, but they will also be given the chance to use any other site they prefer. Order payments will be run from the on-premises private environment, so no personal sensitive data will be transferred to the cloud.

In this design, the central database will be the source of truth for article information, and the on-premises frontend will be the equivalent one for media assets, such as photos and videos of the different collections offered, but inventory and order management will be done at the factory-site level, while payments will be handled on-premises. Updated order and inventory information will be periodically retrieved from the central database too, in order to provide global information about the status of the business.

The APIfied database service running on-premises will return a JSON document containing all articles that are currently part of the different collections offered worldwide, including a description and links to their corresponding media files for each of the products. Factory sites will use this service to update their local data and will combine it with local inventory information to update the list of articles offered in each local online store. A process running on-premises will also periodically poll all factory sites using another APIfied database service located in regional databases that will provide up-to-date information about inventory status and the latest orders.

GKE clusters will be used to scale up and down the number of worker frontend nodes automatically, depending on the level of demand, and a local **content delivery network** (**CDN**) will be used to serve local copies of static media files. The contents of the CDN will also be periodically refreshed during low traffic hours, in order to reduce the number of concurrent requests being sent to on-premises, private frontend systems.

At the network level, this architecture will make use of a **gated topology** to only expose the APIfied database services used to download the latest product catalog from the central database, to obtain the latest updates on inventory and orders from each factory site, and the payments API provided from the on-premises backend. The on-premises private environment should be connected to Google Cloud using **Dedicated Interconnect**.

Let's take a look at a diagram of the suggested architecture:

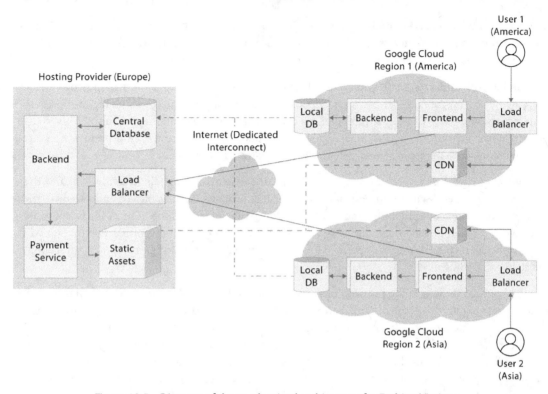

Figure 12.5 – Diagram of the modernized architecture for Fashion Victims

Despite the added complexity of the diagram, opening a new factory would be an easy process. We just need to clone the regional setup to a new region close to the factory, and once the initial cache refresh and catalog synchronization are complete, we would be ready to go.

Hybrid and multi-cloud options

If you remember, we mentioned in other chapters a specific situation that may justify the use of a multi-cloud approach for a scenario such as this one. Imagine that we want to open a new factory in a location that does not have any Google Cloud region close because of legal reasons, for example.

In that case, Fashion Victims may decide to use a local cloud provider and set up a compatible architecture. Since we APIfied the main services, we were abstracting the internals of regional architecture, so as long as the endpoints and formats are the same, we should be able to add a compatible new location. However, we should still sort our connectivity, identity, and security before the new locations are ready to go live and, when possible, try to adapt our tools and processes so that they also work in this new provider. Note that if the alternative provider is AWS or Azure, we can still use Anthos to deploy our clusters, so that can reduce the complexity quite a lot.

Depending on the products and services available in the new provider, building a feature-compatible architecture may be quite challenging, and costs should be considered too before deciding whether to use it or not. The costs to consider should include both development and monthly usage fees. Once we have the numbers, all the pros and cons of using a new provider should be carefully compared against those of using the closest Google Cloud region so that a proper decision can be made. While having a local provider may bring a lot of interesting benefits, as I mentioned earlier, each provider that we add to our equation will exponentially increase its complexity, particularly regarding setup and maintenance tasks, so this is a very important decision to take.

One last piece of advice before we get to the end of this last example: whenever you need to make a decision such as the one just mentioned, don't think only about the present, but also try to think over the medium term; otherwise, you may be wasting time and money for nothing.

And now, it's time to wrap up.

Summary

In this chapter, we examined three very different scenarios and worked together through the process of deciding which design patterns we should use to modernize them and migrate them, either fully or partially, to Google Cloud. We first identified the key areas where we should focus our efforts and which key decisions we needed to take, and defined the right sequence of actions to complete for these migrations to succeed. I hope that deep diving into these real-world scenarios got you better prepared to face your own real-world challenges and use cases with a more intuitive approach.

The next chapter will be the last one in the book, and I can't think of a better way to close it than providing you with a good list of potential migration pitfalls you should watch out for, a few tips that can help you save time and get results faster, and a list of best practices to follow when you are facing a migration to Google Cloud.

Further reading

To learn more about the topics that were covered in this chapter, take a look at the following resources:

- Google Cloud Architecture Center: `https://cloud.google.com/architecture`
- Introduction to BigQuery Omni: `https://cloud.google.com/bigquery/docs/omni-introduction`

13

Migration Pitfalls, Best Practices, and Useful Tips

Congratulations on making it to the last chapter of the book! I thought it would be a good idea to use these last few pages to summarize interesting content that can save you a lot of effort and bad times whenever you need to migrate or redesign an application to run on Google Cloud.

In this chapter, we will first discuss the most common pitfalls that can occur during our daily work as developers. Then, we will look at the process from a different angle and talk about best practices that can help us succeed in our migrations.

Both parts of the chapter will also contain some useful tips that will help you overcome obstacles, successfully face delicate situations, and mitigate the complexity that migrations to the cloud usually entail.

We will cover the following topics in this chapter:

- Common pitfalls while moving to the cloud
- Best practices for bringing your code to Google Cloud

Let's get started!

Common pitfalls while moving to the cloud

Migrating applications and services to Google Cloud is not an easy process. There will be different stages, and pitfalls and challenges may appear in any of them. We should keep these situations in mind even before they happen, so we can quickly mitigate any negative effects they could have on our plans, particularly those that may put our migration schedule at risk.

Let's go through the most common pitfalls of migrating to Google Cloud.

Management not on board

Moving to the cloud will be challenging and working with multiple stakeholders will bring a lot of scrutiny to the table. For this reason, it's imperative that upper management is fully convinced and defends and supports the need for migration.

> **Tip**
>
> Building a solid business case that clearly details all the benefits and provides KPIs to measure the added value of migration to Google Cloud will help you obtain the buy-in from management and will also make it possible to prove such benefits as soon as the earliest wave of migration is completed.

You will find detractors in your way, for sure, so you should be armed with a certain degree of flexibility in your plans and consider everyone's interests while you keep the core of your migration plan as intact as possible. Having people with stakeholder management skills in the team will be extremely beneficial in the early stages of migration.

Unclear strategy and poorly defined migration roadmap

This pitfall is one of the most dangerous ones, and we mentioned it at the beginning of the book. Planning a migration of a single application to Google Cloud is not trivial, so what can I say about moving the whole data center? It will be a complex and tedious process, so it will require a reasonable amount of time to put a good plan together.

As with any other project, we will need a global plan built as a combination of specific migration strategies for each application or service. It will also help us a lot if we have a vision: a clear idea of what we want to achieve. Having this final picture as a tangible objective will help us identify intermediate milestones that we can use in our plan.

We reviewed an interesting list of questions at the end of *Chapter 2, Modern Software Development in Google Cloud*, that we could use to prepare this plan. Some of them cover topics such as which migration approaches to use or which of the **anything-as-a-service** (**XaaS**) models we prefer to use, and we can use them to identify which additional requirements we may have.

It will take us a while to complete the initial analysis and to ensure that we have the right people on board, including all the required skills to perform this assessment. Any gaps detected in this phase should be covered by upskilling existing members, which may not always be possible, or by hiring specialized external consultants.

> **Tip**
>
> Admitting that there are skill gaps in the organization when we are facing a migration and calling out for external help is not a sign of weakness but a sign of common sense and responsibility. Moving into a new technical space comes with its own set of challenges, and having all the required knowledge and skills is often not possible, so hiring external consultants can be a nice option to fill any detected gaps.

Unmanaged expectations regarding benefits and costs

While the benefits of moving to the cloud are easier to explain and will become our best tool to justify the need to modernize our infrastructure and processes, a proper expectation management process should provide all stakeholders with full and transparent information about what a migration to the cloud entails and, particularly, how much it will cost in terms of time and money.

> **Tip**
>
> Estimating the costs of moving our infrastructure to the cloud will be one of the toughest parts of a migration because we will need to take into account both direct and indirect costs, including potential new hires, team upskilling, process updates, and post-migration tasks. This is another reason for having a well-crafted migration strategy plan that will give us a much more accurate action plan that we can later map to time and money.

Mapping the migration of components and waves to their corresponding amounts of time and money will not be possible in all cases, and this may become additional criteria for prioritization since we can identify the best candidates for migration among those that can bring tangible cost savings and additional advantages to our organization since day one of its migration. We can prioritize their migration and use them as early proof of the success of our plan.

Hidden costs will also appear sooner or later once the migration begins. They are very dangerous, and we must keep them to a minimum, mitigating their potential negative effects on our budget.

> **Tip**
>
> Being familiar with the pricing model of the Google Cloud products that our organization will use can help us mitigate some of the hidden costs that may appear during our migration. We can use the Free Tier to accommodate unexpected extra required capacity or even consider sustained use discounts to compensate for the extra costs, sometimes even bringing additional cost savings to our organization. Also, think of your sales representative from Google Cloud as another member of the team, and never stop asking for advice on new opportunities for saving. Finally, keep in mind your organization's potential long-term plans since you can use them to obtain additional savings.

Some commonly incorrect assumptions about migrations to the cloud include believing that lift and shift migrations are immediate or that once an application has been moved to Google Cloud, it no longer requires any additional maintenance or optimization work. Using detailed information for myth-busting inside our organization will help everyone's expectations align with reality and will help us make better decisions.

> **Tip**
>
> Moving to the cloud is a great step forward toward modernizing our infrastructure, but this process will also mean modernizing our organizational culture and mindset, which means changing the way our users conceive and use applications and services. People in our organization will need new skills to properly adapt to these changes, and we will need to include specific action items in our migration plan for this purpose.

Too disruptive or too conservative

We can use different strategies when we are deciding how to approach a migration. A very conservative one would be to only move small and non-critical parts of our infrastructure to the cloud while we keep the core of our business on-premises, just using the cloud as an extension during peak times. We can also be too disruptive, bet it all and move every single of our components to the cloud as soon as possible. Both of these extreme approaches are a bad idea in most cases.

> **Tip**
>
> As with many other decisions in our professional and personal life, being too radical will not often be the best choice, and there are always alternative intermediate approaches that will let our organization enjoy the benefits of Google Cloud from day one while we don't put our eggs in a single basket, and we can progressively increase our presence on the cloud while we still keep some essential services in our private data center.

This is the reason why hybrid and multi-cloud patterns exist and also the reason why we dedicated an important part of this book to talking about and showing you how to use them. Whatever your reason for choosing one of these architectures is, you made a great choice that will increase the available options for hosting your applications and services, will help you avoid vendor lock-in, and will give you a chance to decide later whether you want to become a cloud-native customer, or whether you still prefer to run some of your workloads on-premises.

The good part of these strategies is that you are in control, and if you play the right cards (did anybody say Anthos?), you can run your applications wherever you want. And once you taste the flavor of freedom, you will not like any other.

Not knowing your infrastructure and services well enough

If we want to build a decent migration plan, it is key to have people on the team who are very familiar with our infrastructure, applications, and services in their current state before the migration actually starts. And that's because knowing how they are connected and which dependencies they have can determine the difference between success and failure.

> **Tip**
>
> Having a clear view of application dependencies and a full list of all the components that each service uses will save us a lot of headaches. Of course, a migration process will never be perfect, but we must work hard to make it as easy as we can. Carefully crafted migration waves, containing apps and services that can be moved to Google Cloud together, and having all the information required to prioritize these waves properly will make a huge difference.

Migrating to the cloud is just moving infrastructure

A common pitfall happens when the migration or upper management team underestimates what a migration to Google Cloud entails. While moving infrastructure to the cloud will be a significant part of the migration, we will also be migrating interdependent applications and services which run on this infrastructure, and this will often mean partially modernizing or even fully re-writing their code from scratch.

> **Tip**
>
> Each of our services comes with its own set of tools and processes. This means that modernizing our infrastructure will inevitably involve the modernization of our tools and processes, too. And we should include specific actions in our migration plan to ensure this modernization actually happens.

It's not uncommon to see customers finishing their migrations only to realize that their previous processes are broken or no longer make sense. Thinking about the final picture of our organization that we want to achieve, our vision should include not only the more technical view but also all the processes and the teams that will interact with our shiny new cloud architecture, measuring and improving application performance or applying patches and updates when they are required. Only if we take into account all the details will we be able to identify the different areas that will need their own transformation plan.

And, by the way, if you have not used DevOps and have no SRE team or have never invested in platform engineering, this will be a perfect time to do it. You will not regret it!

But the worst part of all is that we are leaving a key component out of the picture, a common oblivion that may come with a heavy cost: what about our data? This is such an important topic that it will get its very own section.

Being too careless with your data

I mentioned it in a previous chapter: data is a key component of migration, and when we migrate our data, we don't just copy it to a different location. Among some other tasks, depending on each scenario, we will also need to ensure data integrity, define a data backup strategy, assign a data retention plan, and configure data access policies.

> **Tip**
>
> Sometimes moving our data to the cloud will take longer than migrating the infrastructure that will be using it. Identifying dependencies and having a good estimate of the time required to migrate each part of our architecture, where data is just another key component, can help us properly organize and prioritize all our tasks.

Some organizations are extremely careful with where and how they store their data. They keep it safely stored on-premises, and then they set up an even more secure location to store it on their shiny new public cloud provider. Finally, they copy their data to that new provider using blazing-fast FTP, which transfers data without encryption...and this is how the chronicle of a disaster begins.

Our data is, in many cases, our competitive advantage and the result of years of work. And this data has to be protected when being stored, when it's in transit, and even once it had arrived at its destination. I will repeat it just one last time, I promise: security and integrity both follow the principle of the weakest link, so please ensure that none of your links are too weak, or you may be jeopardizing your complete data security strategy.

Even once our data is properly protected in the cloud, we may be following other dubious practices, such as exposing the ports of our databases to the internet. Please try to identify these risks as soon as possible, preferably even before the migration actually happens, and work on mitigating them as part of the plan. In the example of the database, we can expose our services using APIs, executing queries internally, and having all passed parameters checked to prevent SQL injections. If we fix our weak spots, combine them with proper identity, authentication, roles and responsibilities, and foster the use of gated network topologies, we can reduce the risks of intrusions and minimize the opportunities for data theft.

> **Tip**
>
> Optimization is an iterative process, and this is also applicable to security. Our work is never completed, and we may just have completed the latest iteration, but we should start over again shortly and keep on discovering weak spots. Since our organization will constantly be making changes and deploying new services, it makes sense to keep an eye open and keep on trying to identify sources of security risks.

Making migration waves too big or too long

Once we have all the information required to start planning our migration, we will assign each of our components to a different wave.

> **Tip**
>
> A common error when defining migration waves is to make them too big or to put together components that will take too long to migrate. Too much complexity may turn a migration wave into a black hole, and we should prevent this by using an agile-like process, where quick and short sprints get groups of applications migrated to the cloud in a short amount of time. The migration process should also include specific actions to handle issues happening during the migration, and these should be managed in a way that doesn't affect the timelines of other waves, specifically those with no dependencies.

If you find any components that cannot be accommodated into a wave, such as monolithic or extremely complex applications, you should flag them and decide on the best strategy to bring them to the cloud. Sometimes it will make sense to work on their migration once all the sprints are completed, but in other cases, dependencies may force us to migrate them in parallel with our sprints so that the complete migration doesn't take too long.

> **Tip**
>
> A migration to the cloud will take a while, and an organization is like a living being: it's constantly changing and evolving. Our plans should consider that new deployments may be performed during the migration, so we should define a process to minimize the negative effects of these changes. This means that, at some point, new deployments on the legacy infrastructure should no longer be allowed unless they affect components that will remain on premises. We should also do our best to ensure a smooth transition by allowing new deployments on Google Cloud to be performed as soon as possible.

Unbalanced latency, complexity, and cost

When we have users in multiple locations worldwide, sometimes architects get too obsessed with the idea of minimizing latency, often at the risk of increasing complexity and costs, which does not make sense in many scenarios.

The last pitfall to avoid is oversizing for the sake of making a few numbers better while making others much worse. It's good to adapt our architectures to our users, but we should not get lost in endless optimizations that may only benefit a minor portion of our users but have important costs for our organization in terms of money and time or even make things worse for many others of our users.

We may be tempted to think that having replicas of our infrastructure on every continent is always a great idea. Indeed, if our users always get content from a CDN cache located closer to them, they will always have a better experience. While this may be true, we may also be making the mistake of not looking at the whole picture.

> **Tip**
>
> Optimization matters, but costs and complexity also matter, and they usually come together in one pack. Before making any changes to optimize a specific metric, we should carefully study the potential effect of our actions on the rest of the metrics and processes that are key to our organization and weigh the pros and the cons before deciding if it makes sense to go ahead with those changes. This is another reason why knowing our services well and understanding their dependencies is key for any migration to succeed.

If we decide not to use scale to zero in our workloads, this will make the first user in a while have a shorter wait time. But is this worth the extra cost? The answer to this question will be different for each service. If we provide corporate services used by colleagues in our organization during working hours, then probably it won't be required. If we manage micro-transactions that happen constantly, we may not need to enable it either, because resources will always be alive. In some other business scenarios, it may have a beneficial impact on our organization.

I hope you found this list of common pitfalls interesting. I'm sure that if you keep most of them in mind when planning and performing your migration, they can save you quite a few headaches and help you succeed.

And now, let's move on to the next section, which includes some best practices that will also help us a lot in our migrations.

Best practices for bringing your code to Google Cloud

While no migration to the cloud is exempt from risks, anticipating what could go wrong can help us have countermeasures ready to be implemented, thus reducing the negative effects, and that's the main reason why we went through the list of common pitfalls.

But another way of making our migration less troublesome is to follow best practices that will also help us mitigate those risks, but this time by following specific actions and guidelines.

Either if we are designing a cloud-native application or if we are modernizing a legacy one to bring it to Google Cloud, we should keep in mind the following list of recommendations.

Avoid migrate and delete

I already mentioned this at some point in the book, and this is a common mistake that repeats again and again. Services are often migrated to the cloud just to have them decommissioned a few weeks later. This is an absolute waste of time and can have a negative effect on our cloud migration team.

We should take our time to understand our architecture, study usage patterns and metrics, and then proceed to flag which components and services are no longer needed. And if no policies or regulatory requirements prevent us from doing it, we should then shut them down and consider them out of the scope of our migration.

If once our migration is complete, which should happen a few months later, nobody complained because they were offline, my recommendation is to keep their latest backups, so we can restore the services later if there is a very good reason for doing it, and we can safely proceed to decommission all the components as soon as possible.

> **Tip**
>
> There are companies that are considering a migration to the cloud, but they have no application and service usage data at all. If you happen to work for one of them, it would be a great idea to start collecting this valuable information for a few weeks while you start the planning phase of the migration. The money and time that you will need to implement this collection process will be nothing compared to the benefits you will obtain from the collected data during the planning phase of the migration, and you could also use the opportunity to include this good practice in the core of your future cloud architecture.

We can save a lot of time and computing resources with this simple tip, and this will also avoid bringing additional fatigue to our migration team, which should be focused on migrating critical services for our organization and not wasting time with useless tasks.

Check regulations and compliance before moving ahead

This was also mentioned in earlier chapters, and it can be a source of trouble if it's not considered in every migration to the cloud. Please ensure that moving your applications, and particularly their associated data, will not have any regulatory or compliance implications that you may regret later.

Be particularly careful with **Personally Identifiable Data (PII)**, especially if you are planning to host your cloud workloads in a different location than your premises. Moving to a different country or even to a different continent, will exponentially increase the risks of unwillingly breaking any law or regulation, so having the legal team on board before taking the go/no-go decision for the migration will always be an excellent idea.

Also make sure to ask which certifications your cloud provider has because these can be a good aid to prove that they have your back. For example, you can see the list of Google Cloud compliance offerings by visiting this link: `https://cloud.google.com/security/compliance/offerings`.

> **Tip**
>
> Remember that hybrid and multi-cloud patterns exist not only to avoid vendor lock-in but also for situations like this, and not being able to migrate key services to the cloud because of a compliance issue shouldn't be a blocker to migrating the rest of your infrastructure. Also, please remember that the different available public cloud providers are often opening new regions worldwide, so your blocking may be only temporary, and you should keep on checking for updates regularly because there may be alternatives to move the affected services in the near future.

And let me insist on one other related key topic: please make sure that your migration team is not only formed of technical profiles but also by people with expertise in other areas who can provide support from the different angles that migration entails: from compliance experts to architects, network and security experts, system administrators, developers (of course!) and external consultants, when required.

Prepare a migration toolkit in advance

While many organizations want to start migrating applications and infrastructure to the cloud as soon as possible, the first technical implementation in cloud migration, once we have properly configured our cloud account and project, should be defining a set of tools and processes which work across all our different providers and environments.

We should have a complete toolkit in place, including CI/CD and a proper monitoring and logging system offering centralized observability, after implementing the principle of least security and a zero-trust model. These should be combined with a set of IT and operations processes compatible with our hybrid or multi-cloud architectures. And only when these elements are in place we will be ready to start planning our migration waves.

> **Tip**
>
> Some tools can help us automate many of our processes in Google Cloud to help us save a lot of time and increase the reliability of our deployments. A good example is Terraform (`https://cloud.google.com/docs/terraform`). You can find a link in the *Further reading* section of this chapter to a book also published by Packt Publishing about this amazing tool and how to use Terraform to build a strong Google Cloud foundation. I had the privilege of technically reviewing that book and can say that Patrick Haggerty did an amazing job explaining how to set up our Google Cloud projects in the best possible way from every possible angle.

Try to simplify, that's what the cloud is about

Each organization may have its own reasons for moving to Google Cloud, but in general, users are looking for elastic and almost infinite additional and easy-to-use computing resources, a reasonable pay-per-use pricing model, and very short delivery times. In summary, we love how simple and quick everything is on the cloud. So, please, don't ruin this beautiful picture with some ancient and dreary corporate processes.

I have seen automatic delivery workflows, which can be executed end to end in a couple of minutes, paired with approval processes that take weeks to complete, and I'm sure you also have your own hall of shame with many other examples. Trying to reuse the same processes that we defined years ago and used on premises in our cloud environments will be a nonsense in most cases. Power and speed are two key features of Google Cloud, and our processes should not only not harm them but also contribute to maximizing their beneficial effects on our organization.

> **Tip**
>
> Try to make your processes as simple as possible. This doesn't mean giving up on security, control, or any other desirable features, but just making sure that all the benefits of the XaaS model that we decided to adopt in the cloud, delegating part of our work to our platform, are also reflected on our processes by removing all those steps that no longer make sense. Don't waste time with additional bureaucracy or approvals for components that can be requested with just one click.
>
> You should better use available products and features, such as roles and permissions, to delimit who can do what and implement centralized observability and controls of cloud usage and its associated costs. This way, you can let your users thrive and enjoy all the amazing features that Google Cloud provides while you have access to real-time data and insights about what's going on in the platform.

Analyze and prioritize accordingly

Spending a reasonable amount of time performing an initial analysis at the beginning of our migration is a decision that we will never regret. It may be frustrating for many of our stakeholders to see the technical work being delayed, but this stage of the migration must take its time to be properly completed. Being able to correctly prioritize which applications and services should be migrated first will save us a lot of time, money, and headaches and can also be used to boost the morale of the migration team and to share the quick wins with the rest of our organization and also use them to propitiate a cloud-friendly culture.

> **Tip**
>
> If any applications are clearly identified as cloud-native, they can be very good candidates for the earliest migration wave. Developers can work in parallel on those other applications which can be modernized so that they are ready to be migrated as soon as possible. And we can decide what to do with the remaining ones, using one of the cloud migration approaches that we discussed earlier in the book: either split them into pieces, use "lift and shift," replace them with a more modern functionally compatible alternative, or re-write them from scratch to make them cloud native. Don't forget to always consider the pros and cons of each option, including how much it will cost in terms of both time and money, and always take into account whether there are any dependencies that may also affect additional applications and services.

Migration to the cloud is an amazing chance for us to make things simpler: split monoliths, reduce risks, interconnect and reuse where possible, and use the opportunity to modernize our whole organization and its culture.

Measure, analyze, optimize, and prove success

While our migration may still be ongoing, and even once it's finished, our work will not be done yet. As part of our new strategy in the cloud, we should have implemented performance and availability monitoring and metric collection among our key management processes on Google Cloud.

Observability should be one of our key priorities to guarantee the success of our migration strategy. We should constantly measure application performance and its associated key metrics to identify potential points of failure, performance bottlenecks, and areas of improvement.

> **Tip**
>
> We should never forget that our Google Cloud environment will be dynamic, so we should repeat our analysis of opportunities for improvement periodically so that we can also consider the effect of new deployments and changes that happened since our last analysis. We should also be aware of all the announcements made by Google, including not only notifications about new services, improved features, and technical upgrades but also any news on price updates and discounts that we could benefit from, especially if the number of cloud resources that we are using increases and we move to a different tier.

We can use our experience during the migration process to extend our original list of KPIs with additional information and use all the collected data to prove the success of our migration and offer our stakeholders a clear view of the benefits and changes that Google Cloud has already brought to our organization.

> **Tip**
>
> While we have already discussed the importance of putting together a team with different skills and profiles to safely face our migration to the cloud, it's also important to think about which roles will be needed once the process is completed. Analysts and cloud security experts could be examples of these new roles, but we should also consider having architects and project managers who are in touch with Google Cloud account managers and experts so that new opportunities can be identified, including participation in betas, early notification of new features, access to learning and certification programs or access to additional discounts with a committed-use contract.

Connectivity and latency can make a huge difference

Most migrations involve a hybrid scenario, sometimes permanent and others just temporary until all our workloads have been moved to the cloud. In all these cases, and especially for those making use of long-term hybrid and multi-cloud architectures, combining the right network topologies and getting connected using an option that offers low latency and decent bandwidth will be key factors in shaping the user experience of our new environment.

A common fear users express when they are new to the cloud is that everything becomes slower due to distance and latency. While a bigger distance between our computer and any servers we connect to may be associated with higher latency, this may not be true in most cases, since even when we work on-premises, we may suffer networking issues. Everything will depend on the quality of our connection, and this is where networking experts can help us choose among the many connectivity options and network design patterns that are available. They should also carefully analyze where redundant connections will be required to provide backup paths for our traffic. These best practices will guarantee a good user experience and a reliable topology that can handle temporary failures in some of our connections and links without causing downtime in our services.

> **Tip**
>
> Since our organization may have a presence in multiple regions, it's important to have a clear picture of who will be using and who will be managing our applications and where they will be connecting from to do it. The answers to these questions will be used to decide which is the best connectivity option to use and which Google Cloud region or regions we should deploy our services. And never forget that organizational needs may change, so we will need to periodically check whether the answers that we got in the past are still valid or whether we need to make upgrades to our original setup.

While all these decisions will not be written in stone, it's important to make the best possible choices by studying all the information and metrics that we have available so that we can provide the best experience to our users.

This is an important topic that can directly affect cloud adoption because if we are able to provide a good experience to our users and administrators, our organization will not see noticeable differences in user experience after moving to the cloud. The added features that Google Cloud can also provide will be perceived as a huge leap forward.

> **Tip**
>
> We didn't dive too deep into the networking part of Google Cloud in the book since it is targeted at developers, but networking is a key area in any migration to the cloud. Having network experts among the team responsible for the migration plan first, and for cloud operations, once the migration is completed will be a must if we want our migration to Google Cloud to succeed.

Build once, deploy anywhere, and as many times as required

I want to use this last best practice to summarize some of the key messages of this book on how to build applications for Google Cloud.

> **Tip**
>
> Using self-contained architectures with workloads that can be deployed in more than one cloud provider and in any of their regions will help us respond much faster to any organizational needs, such as moving services to other regions and zones or adding extra replicas in new locations to improve the user experience of local customers.

Since migrating to the cloud implies taking a lot of decisions, we will make the wrong ones at some point despite all our efforts to avoid them. It will just happen, and we should accept the inevitable but also be well prepared to act when it happens. And for this reason, it's important to use design and development patterns that can properly handle failures and that we can quickly deploy somewhere else if we suddenly need to make corrections or improvements to our architecture.

And being able to deploy anywhere any time, and as many times as we want means that we have total freedom and control over where our code will run and, personally, I think this is one of the best situations in which we, as developers, can see ourselves, so we shouldn't underestimate the opportunity and do our best to bring it to our organizations.

And this was all regarding best practices. I hope you found all this information useful and interesting. Now, it's time for the final wrap-up.

Summary

We started the last chapter of the book by identifying the most common pitfalls that happen while we move or modernize our applications and services to benefit from the multiple advantages that Google Cloud offers.

Then, we discussed some of the best practices recommendable when bringing our code to Google Cloud.

Both parts of the chapter also included some tips to help you overcome obstacles, handle delicate situations that are quite usual in this kind of migration, and mitigate the complexity of this kind of process.

And we got to the end of the chapter! Thank you so much for finishing the book, and I hope that it helped you better understand how you can start making the most out of Google Cloud as a developer.

You chose an amazing technical area to develop your skills, where there are still a lot of new applications and breakthroughs to be ideated and offered, and I'm sure you will be an important part of this digital revolution. I would love it if you let me know which amazing projects you worked on after reading this book, and I wish you the best of luck on your journey with Google Cloud.

Further reading

To learn more about the topics that were covered in this chapter, take a look at the following resources:

- *The Ultimate Guide to Building a Google Cloud Foundation*: `https://www.packtpub.com/product/the-ultimate-guide-to-building-a-google-cloud-foundation/9781803240855`
- *Visualizing Google Cloud*: `https://cloud.google.com/blog/topics/developers-practitioners/introducing-visualizing google-cloud-101-illustrated-references-cloud-engineers-and-architects`

Index

Packtpub.com

Subscribe to our online digital library for full access to over 7,000 books and videos, as well as industry leading tools to help you plan your personal development and advance your career. For more information, please visit our website.

Why subscribe?

- Spend less time learning and more time coding with practical eBooks and Videos from over 4,000 industry professionals

- Improve your learning with Skill Plans built especially for you

- Get a free eBook or video every month

- Fully searchable for easy access to vital information

- Copy and paste, print, and bookmark content

Did you know that Packt offers eBook versions of every book published, with PDF and ePub files available? You can upgrade to the eBook version at packtpub.com and as a print book customer, you are entitled to a discount on the eBook copy. Get in touch with us at customercare@packtpub.com for more details.

At www.packtpub.com, you can also read a collection of free technical articles, sign up for a range of free newsletters, and receive exclusive discounts and offers on Packt books and eBooks.

Other Books You May Enjoy

If you enjoyed this book, you may be interested in these other books by Packt:

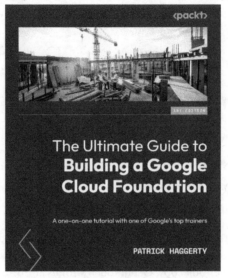

The Ultimate Guide to Building a Google Cloud Foundation

Patrick Haggerty

ISBN: 978-1-80324-085-5

- Create an organizational resource hierarchy in Google Cloud
- Configure user access, permissions, and key Google Cloud Platform (GCP) security groups
- Construct well thought out, scalable, and secure virtual networks
- Stay informed about the latest logging and monitoring best practices
- Leverage Terraform infrastructure as code automation to eliminate toil
- Limit access with IAM policy bindings and organizational policies
- Implement Google's secure foundation blueprint

Terraform for Google Cloud Essential Guide

Bernd Nordhausen

ISBN: 978-1-80461-962-9

- Authenticate Terraform in Google Cloud using multiple methods
- Write efficient Terraform code
- Use modules to share Terraform templates
- Manage multiple environments in Google Cloud
- Apply Terraform to deploy multi-tiered architectures
- Use public modules to deploy complex architectures quickly
- Integrate Terraform into your Google Cloud environment

Packt is searching for authors like you

If you're interested in becoming an author for Packt, please visit authors.packtpub.com and apply today. We have worked with thousands of developers and tech professionals, just like you, to help them share their insight with the global tech community. You can make a general application, apply for a specific hot topic that we are recruiting an author for, or submit your own idea.

Share Your Thoughts

Now you've finished *Google Cloud for Developers*, we'd love to hear your thoughts! Scan the QR code below to go straight to the Amazon review page for this book and share your feedback or leave a review on the site that you purchased it from.

https://packt.link/r/1-837-63074-7

Your review is important to us and the tech community and will help us make sure we're delivering excellent quality content.

Download a free PDF copy of this book

Thanks for purchasing this book!

Do you like to read on the go but are unable to carry your print books everywhere?

Is your eBook purchase not compatible with the device of your choice?

Don't worry, now with every Packt book you get a DRM-free PDF version of that book at no cost.

Read anywhere, any place, on any device. Search, copy, and paste code from your favorite technical books directly into your application.

The perks don't stop there, you can get exclusive access to discounts, newsletters, and great free content in your inbox daily

Follow these simple steps to get the benefits:

1. Scan the QR code or visit the link below

https://packt.link/free-ebook/9781837630745

2. Submit your proof of purchase
3. That's it! We'll send your free PDF and other benefits to your email directly

www.ingramcontent.com/pod-product-compliance
Lightning Source LLC
LaVergne TN
LVHW062303060326
832902LV00013B/2031